Advances in Anatomy
Embryology and Cell Biology

To all Subscribers

Please Note:

With this volume of Advances in Anatomy, Embryology and Cell Biology, the format and appearance of the series will change. All forthcoming titles will be published as individual volumes; the subdivision into partial volumes will be discontinued. This means that this title (Volume 56) is the direct successor of Vol. 55/5. There will be <u>no</u> Vol. 55/6.

Subscriptions will not be affected in any way by this change.

In future only the English title ,,Advances in Anatomy, Embryology and Cell Biology" will be displayed.

The Publisher

Springer-Verlag Berlin Heidelberg New York

Advances in Anatomy
Embryology and Cell Biology

Vol. 56

Brigitte Kaissling
Wilhelm Kriz

Structural Analysis of the Rabbit Kidney

With 47 Figures

Springer-Verlag
Berlin Heidelberg New York 1979

Dr. Brigitte Kaissling, Prof. Dr. Wilhelm Kriz, Anatomisches
Institut der Universität Heidelberg, Im Neuenheimer Feld 307
D-6900 Heidelberg
With the cooperation of: Dr. Lise Bankir, Hôpital Necker,
Paris, Prof. Dr. Joseph M. Barrett, Medical College of Georgia,
Augusta, USA, Dr. Hermann Koepsell, Max-Planck-Institut für
Biophysik, Frankfurt/Main
Technical Assistance: Saliha Šabanović, Anatomisches Institut,
Heidelberg

ISBN-13: 978-3-540-09145-5   e-ISBN-13: 978-3-642-67147-0
DOI: 10.1007/978-3-642-67147-0

Library of Congress Cataloging in Publication Data. Kaissling, Brigitte,
1942- Structural analysis of the rabbit kidney. (Advances in anatomy, embryology, and cell biology; v. 56) Bibliography: p. Includes index. 1. Kidneys. 2. Histology. 3. Rabbits—Anatomy. I. Kriz, Wilhelm, 1936- joint
author. II. Title. III. Series. QL801.E67 vol. 56 [QL873] 574.4'08s
[599'.322] 78-25810

Composition: SatzStudio Pfeifer, Germering

2121/3321-543210

Dedicated to Professor Otto Bucher (Lausanne)
on the occasion of his 65th birthday

Contents

1 Introduction

Karl Peter stated in 1909 in his famous monograph, *Untersuchungen über Bau und Entwicklung der Niere,* that the rabbit has a simply structured kidney and without knowledge of it he would not have succeeded in understanding the structural organization of other more complicated kidneys. Even with today's advanced knowledge of kidney organization this statement is still valid. With regard to the histotopographic relations of tubules and vessels as well as to the ultrastructure, the rabbit kidney is more simply organized than other mammalian kidneys, for example the mouse kidney and especially the Psammomys kidney. The rabbit kidney essentially displays the basic pattern of a mammalian kidney. Therefore, a thorough knowledge of the morphology of the rabbit kidney is indispensable for investigating kidneys of other species.

Another reason for extensively investigating the rabbit kidney is that it is the only kidney with tubules that can easily be isolated and, therefore, be used for in vitro experiments (Burg et al., 1966). Many investigations using these techniques have been performed in the last decade (for references see corresponding chapters). Accordingly the different segments of the nephron are functionally best characterized in the rabbit kidney. The purpose of this paper is to supply the basis for good structural-functional correlations.

2 Material and Methods

2.1 Light Microscopy

The histologic and histotopographic observations are based on a study of the kidneys of a total of 25 male and female New Zealand rabbits (2500-4500 g body wt.).

2.1.1 General Histology

Following anesthesia with Nembutal (20 mg/kg body wt. I. P.), seven female rabbits were tracheotomized and submitted to osmotic diuresis by infusion of 50 ml of a 20% mannitol solution at a rate of 0.5 ml/min into the left jugular vein.

The kidneys were surgically exposed and rapidly removed after clamping the renal pedicles and then fixed in toto in Bouin's solution. In three animals the renal vein was ligated for 30 prior to clamping the total renal pedicle and removing the total kidney.

After dehydration and embedding in Paraplast, the kidneys were cut in 8-μm-thick serial sections, which were alternately stained with the Azan- and the PAS-techniques.

In addition, Epon- embedded tissue (see 2.2) was used for light microscopic studies. The material was cut in 1-μm thick serial sections on a Sorvall MT II B ultratome and stained with Azur II and Methylene blue (Richardson et al., 1960).

2.1.2 Single Nephron Injections

Following anesthesia with Nembutal, tracheotomy, and osmotic diuresis (see 2.1.1), the kidneys of nine female rabbits (body wt. ~ 2500 g) were cleaned of blood with Haemaceel by retrograde perfusion via the abdominal aorta. They were subsequently perfused with a 1.4% glutaraldehyde/ 0.1 *M* phosphate buffer solution. After removal, the total kidneys were immersed in the same fixative solution for at least 12h After being washed in buffer solution, they were stored in 70% alcohol.

Single nephron injections were carried out on fixed kidneys using uncatalyzed silicone rubber (Microfil). Short loops of Henle were filled from the kidney surface; long loops of Henle, from a cross-sectional plane trough the outer stripe as described by Kriz et al. (1972a). Embedded in Paraplast the kidneys were cut in 10-μm-thick serial sections, which were stained with hematoxylin -eosin. Twenty well-injected nephrons with short loops and 15 with long loops were traced in the serial sections.

2.1.3 Vascular Injections

In six New Zealand rabbits (body wt. ~3000 g) of both sexes the blood vessels were filled with silicone rubber (Microfil). Following anesthesia with Nembutal (see 2.1.1), the abdominal aorta, the inferior vena cava, and the renal veins were surgically exposed.

Arterial injections. The aorta was retrogradely cannulated and Microfil (1 vol MV 111 White and 1 vol diluent and 5% curing agent) was slowly injected by hand via a catheter into the aorta over a span of a few minutes. Immediately after the injection began, the aorta was clamped just above the renal arteries and the vena cava was opened to permit blood and subsequently Microfil to pour out. The injection pressure was not measured, but the kidneys were observed with a binocular microscope throughout the filling procedure. No sign of capillary rupture was detected on the kidney surface.

Venous injections. The renal veins were catheterized after ligation of the aorta above the renal arteries. Microfil was injected by hand under gentle pressure for a few minutes.

The kidneys were removed 1 h after the injection, when Microfil had solidified. They were dehydrated in graded series of alcohols and cleared by immersion in methyl salicylate. The kidneys were then embedded in soft paraffin and 300-μm-thick sections were cut in a plane either parallel or perpendicular to the axis of the vasa recta. The sections were mounted and photographed with a photomacroscope (Wild-Leitz).

2.2 Electron Microscopy

The kidneys of nine male New Zealand rabbits (body wt. 3000-4500 g) were prepared for electron microscopy investigations. The animals were deprived of food for at least 24 h prior to perfusion but had free access to water. To minimize stress on the animals, they were given a premedication of Dehydrobenzperidol (droperidol; 5 mg/kg body wt.; I. M.) 15-20 min. before anesthesia with Nembutal (20 mg/kg body wt.; I. P.). The abdominal aorta was then surgically exposed and retrogradely cannulated. It was clamped above the renal arteries after perfusion started and the inferior vena cava was opened.

For washing-out the blood an oxygenated Ringer solution was used, added with $CaCl_2 \cdot H_2O$ (0.33 g/liter), procain- HCL (5 g/liter), PVP (25 g/liter), and Liquemin (1 ml/liter corresponding to 5000 USP- U Heparin/liter); osmolality ~ 350 mosmol; pH 7.3; temperature 37°C; perfusion pressure ~140 Hg.

Immediately after washing out the blood, perfusion with the fixative solution was carried out without pressure drop. During the course of investigation, several fixatives were used:

a) 3% glutaraldehyde in 0.1 M cacodylate buffer, added with 0.66 g/liter $CaCl_2 \cdot 2 H_2O$; osmolality ~500 mosmol; pH 7.3.

b) 3% glutaraldehyde in 0.1 M cacodylate buffer, added with 0.66 g/liter $CaCl_2 \cdot 2 H_2O$ and picric acid 0.5 g/liter; osmolality ~550 mosmol; pH 7.3.

c) 3% glutaraldehyde + 3% formaldehyde in 0.1 M cacodylate buffer, added with 0.66 g/liter $CaCl_2 \cdot 2 H_2O$ and picric acid 0.5 g/liter; osmolality ~1700; pH 7.3.

The best results were obtained with solution b; the addition of formaldehyde did not obviously improve the fixation quality.

Every animal was perfused with 2000 ml fixative solution at a pressure of 140 mm Hg and a temperature of 37°C. After perfusion fixation, the kidneys were removed and cut in slices that were immersed in the fixative solution for at least 12 h at 4°C. The tissue was then thoroughly washed in 0.1 M cacodylate buffer (pH 7.3; osmolatity increased with sucrose to about 350 mosmol), and cut in pieces of 2-mm-side length and 0.5-mm thickness. From every kidney zone, blocks were chosen which were oriented either transversely or longitudinally to the axis of the medullary rays. Postfixation (maximally 2 h) was carried out with osmium tetroxide in 0.1 M cacodylate buffer. Some blocks were additionally immersed in a solution of 1% uranylacetate in maleate buffer, pH 6.

After dehydration in graded series of alcohols and propylene oxide, the tissue blocks were embedded in Epon. Ultrathin sections were cut on a Sorvall MT II B or a LKB IV ultratome. The sections were stained with lead citrate and uranylacetate and examined in a Philips 301 electron microscope or a Philips 301 with goniometerstage.

The apical-basal depth of the tight junctions was measured on the EM−plane film negatives (magnification x 59,000 or 71,000) in which the junctions were well discernible over their entire depth. Since it is impossible to judge if the zonula occludens in TEM sections has been cut exactly perpendicular, the measured values may be slightly high. On the other hand, in the case of a large scatter very deep junctions may have been considered not perpendicular and dismissed. Therefore the values are considered to be approximate.

3 Structural Organization

The following chapter offers a description of microscopic structure and architecture as well as a comparative and functional judgement of structural organization. The findings of our investigations will be combined with data from literature relevant to this topic.

3.1 Renal Pelvis and General Description

The rabbit kidney is a unipapillary kidney. As very plastically described by Sheehan and Davis (1959), the medulla of the rabbit kidney is shaped like a segment of an organe with the narrow ends curled round the upper and lower poles of the kidney so that they point towards the hilus. The papilla itself is a projection of the central part of the medulla. It is shaped like a cone that has been considerably flattened from front to back. The structures of the medulla (loops of Henle, collecting ducts, vasa recta) converge from all sides toward the papilla; those from the ventral and dorsal parts of the kidney extend in arcs around the outside of the pelvis and are massed into about four or five "peripelvic columns" on each side of the pelvis (Figs. 1 and 2) (Sheehan and Davis, 1959).

The renal pelvis (Sheehan and Davis, 1959; Fourman and Moffat, 1971) forms two septa, which develop from the tissue filling the "hilar tunnel" and which enclose the medullary papilla. The lumen surrounding the papilla and laterally bordered by the pelvic septa is called the primary pelvic cavity. From each main septum about three or four subsidiary septa arise. They contain the interlobar vessels and finally penetrate into the renal parenchyma. Between these subsidiary septa the pelvic cavity forms secondary pouches that penetrate back toward the hilus between the outer surface of the main pelvic septa and the inner surface of the peripelvic columns (Figs. 1 and 2). The primary pelvic cavity and the secondary pouches communicate only at the fornices over the free semi-lunar edges of the main septa; the secondary pouches are separated from each other by the subsidiary septa (Fig. 2).

When the interlobar/arcuate vessels between the peripelvic columns (lying within the subsidiary septa) finally penetrate into the renal parenchyma, they directly reach

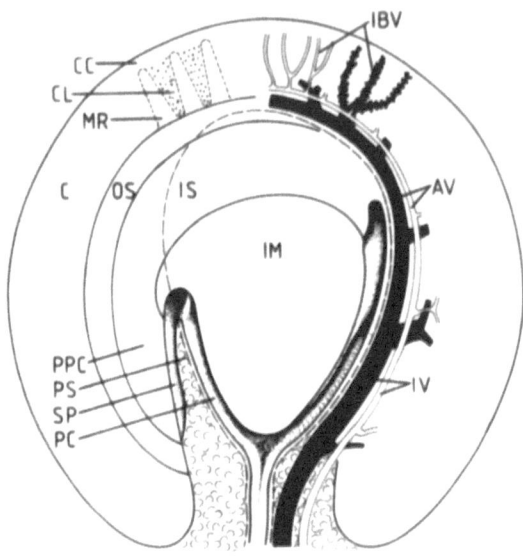

Fig. 1. Longitudinal middle section, demonstrating the general structural organization of the kidney and the renal pelvis. C = cortex; CC = cortex corticis; CL = cortical labyrinth; MR = medullary ray; OS = outer stripe; IS = inner stripe; IM = inner medulla; PS = pelvic septum; PC = pelvic cavity; SP = secondary pouch; PPC = peripelvic column; $IV/AV/IBV$ = interlobular, arcuate, and interlobular vessels (veins are drawn black). The *broken line* indicates the connective tissue anchoring the pelvic septa in the kidney

the corticomedullary border. They carry with them bands of fibrous connective tissue that, lying on the medullary side of the vessels, form fibrous arcs within the renal parenchyma. The arcs may serve as a connective tissue skeleton to which the pelvic septa are anchored (Fig. 1) (Sheehan and Davis, 1959; Doležel, 1975).

In histologic sections the different regions of the kidney are easily discernible. The cortex (Fig. 1) consists of the cortical labyrinth and the slender medullary rays. The cortex corticis, devoid of glomeruli, is well developed; it covers the tops of the medullary rays. The most basal parts of the cortical labyrinth constitute the border toward the medulla; at the transition from the medullary rays to the outer stripe the cortico-medullary border cannot be exactly delineated.

The medulla (Fig. 1) is divided into an outer zone (outer medulla) which again is subdivided into an outer stripe and an inner stripe, and an inner zone (inner medulla), which forms the papilla. The free surfaces of the inner zone (i.e., the papilla) shut upon the primary pelvis cavity; the free surfaces of the inner stripe (i.e., the peripelvis columns) shut upon the lumen of the secondary pouches. The transition from the inner stripe to the inner zone occurs at the fornices. The outer stripe as well as the cortex never directly face the renal cavity.

The thickness of the renal zones has been measured in strict longitudinal, perfusion-fixed, unembedded sections as well as in paraffin-embedded kidneys. The paraffin-

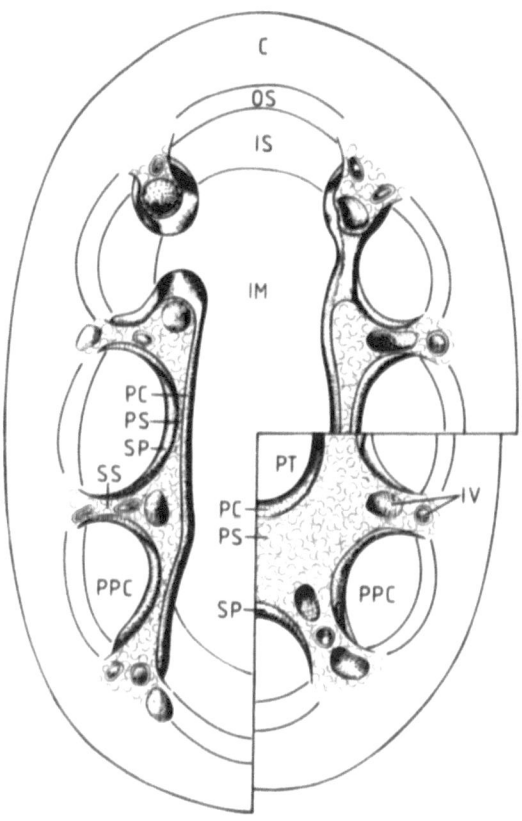

Fig. 2. Transverse section through the kidney at about the middle level of the inner zone. A deeper section through the tip of the papilla is represented in the lower right quarter. The relationships between the pelvis cavities, the pelvis septa, and the kidney are shown. *PT* = papillary tip, *SS* = subsidiary septum; all other abbreviations as in Fig. 1

embedded material, especially the papilla, is shrunken. In the unembedded material the borders between cortex and outer stripe, as well as between outer and inner stripe, cannot be exactly determined. Therefore, the exactly measurable values from the unembedded material were taken; i.e. the distance from the renal surface to the border between the inner and outer zones as well as the distance from this border to the tip of the papilla. The former distance was then subdivided according to the relations derived from the embedded material. The cortex, including the cortex corticis (~0.4 mm thick), is approximately 3.3 mm thick. The outer medullary zone, consisting of the 1.1 mm outer stripe and 2.3 mm inner stripe, measures about 3.4 mm thickness.The inner zone(reaching from the junction of inner and outer zone to the tip of the papilla) has a thickness of 9.0—9.5 mm. These values agree with those obtained by Peter (1909). Sperber (1944) reports considerably higher values, which may be due to differences in body weight; however, the interrelations of the renal zones correspond.

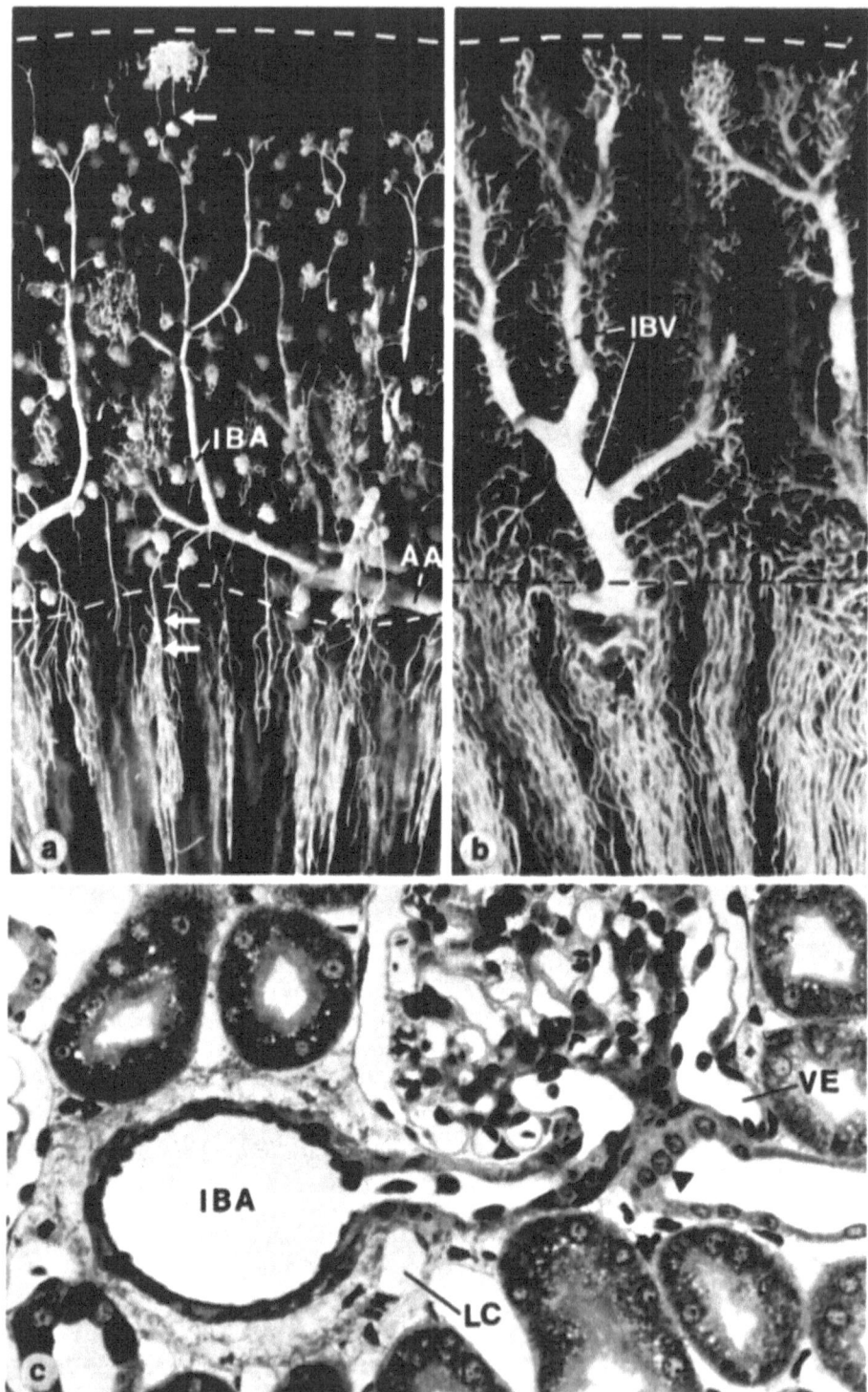

Fig. 3

3.2. Blood Vessels

(In cooperation with L. Bankir, Hôpital Necker, Paris, France.)

The most thorough investigations of the vasculature of the rabbit kidney have been conducted by Trueta and co-workers (1947), more recently by Fourman and Moffat (1971) and by Bankir (1973, 1976 a, b). The vessels in rabbit kidney almost perfectly represent what might be called the basic vascular pattern of the mammalian kidney (Kriz et al., 1976).

The renal artery branches off into eight-ten interlobar arteries, which ascend within the subsidiary septa of the pelvis (Fig. 1) and finally penetrate the renal parenchyma. Thereby, they as well as their branches gradually change their direction and follow an arclike course at the border between the renal cortex and the renal medulla. They are called arcuate arteries, although in the rabbit they do not form arches (on the contrary, the arcuate veins form anatomosing arcs). Therefore, like all arteries in the kidney, they must be considered terminal. The interlobular arteries arise from the arcuate arteries and radially penetrate the renal cortex, ending not far beneath the uppermost layer of renal corpuscles (Figs. 3 a and 5 a).

The afferent arterioles (Figs. 3a and 5a) arising typically at different angles from the interlobular arteries and in part from arcuate arteries, supply the glomerular tufts of the renal corpuscle; they are drained by the efferent arterioles.

Three different types of renal corpuscles can be distinguished (Bankir, 1973): superficial ($\sim$28%), juxtamedullary ($\sim$9%), and the rest, midcortical renal corpuscles ($\sim$63%). The superficial corpuscles are characterized by their vasa efferentia, which run directly through the cortex corticis to the renal surface (Fig. 3 a). Although they often give rise to capillaries already within the cortex corticis, they ultimately split off into capillaries at the renal surface. The juxtamedullary renal corpuscles give rise to the arterial vasa recta (Fig. 3 a) and, therefore, are the supplying vessels of the renal medulla. The bulk of the midcortical renal corpuscles belongs neither to the superficial nor to the juxtamedullary group; they have relatively short vasa efferentia.

The superficial as well as the midcortical renal corpuscles are smaller in diameter than the juxtamedullary corpuscles (Peter, 1909: 120-132 μm in the first group versus 124-172 μm in juxtamedullary corpuscles, Bankir and Farman (1973) measured in injected specimens: 154 μm versus 185 μm; our data – measured in sections of epon-embedded material: 130 μm versus 150 μm). An even more pronounced difference exists in the diameter of the efferent arterioles (Bankir and Farman, 1973): in juxtamedullary renal corpuscles the diameter of the vasa efferentia is more

---

Fig. 3 a c. Cortical vessels. *a.* Arterial vessels filled with microfil. The arcuate artery *(AA)* passes over into an interlobular artery *(IBA)*. The vasa efferentia of superficial glomeruli ascend in a straight manner to the renal surface *(arrow)* and the vasa efferentia of juxtamedullary glomeruli establish the vascular bundles *(double arrow).* x $\sim$ 22. *b.* Venous vessels filled with microfil. The interlobular veins *(IBV)* start just beneath the renal surface. Their basal parts (together with the arcuate veins) accept the medullary venous vasa recta. *Broken lines* indicate the renal surface and the approximate location of the corticomedullary border. x $\sim$ 22. *c.* Semithin cross section of an interlobular artery *(IBA)* giving rise to a midcortical vas afferens. The interlobular artery is surrounded by loose connective tissue in which a lymphatic capillary *(LC;* verified by electron microscopy) is embedded. Macula densa *(arrow head); VE* = vas efferens. x $\sim$390

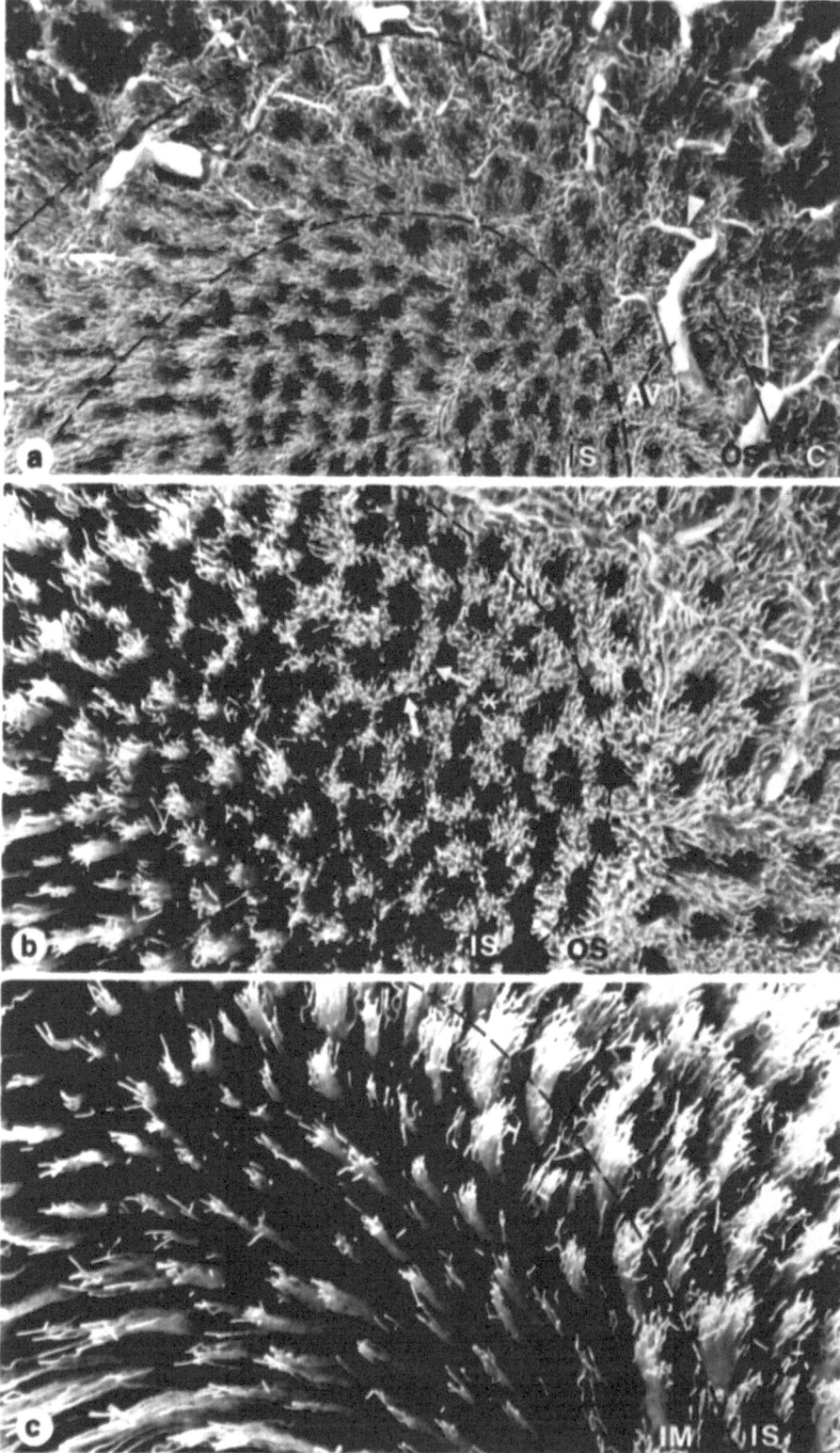

Fig. 4

8

than twice as thick as in superficial and midcortical corpuscles (about 28 μm versus 12 μm); thus, it is thicker than the diameter of the afferent arteriole ($\sim$20 μm in all three types of renal corpuscles).

The vasa efferentia of the superficial glomeruli supply the cortex corticis and, thus, the convoluted tubules of the superficial nephrons (see later). Although considerable overlapping occurs and convoluted tubules from deeper nephrons may also touch the renal surface, it may generally be stated as regards superficial nephrons that the convoluted tubules of a nephron are supplied by capillaries derived from their own vas efferens.

After a small unbranched part the midcortical vasa efferentia split off into the capillaries of the renal cortex forming two different but continuous plexuses: namely, the dense round-meshed plexus of the cortical labyrinth and the less dense longmeshed plexus of the medullary rays. These plexuses are continuous with cortex corticis capillaries.

From these three cortical capillary plexuses the blood is drained into the interlobular veins (Figs. 3 b and 5 a), which start very superficially in the cortex corticis and leave the cortex via the arterial route. It is worthwhile mentioning that blood from the medullary ray plexus has to pass through the labyrinth capillaries to gain access to the interlobular veins, whereas the labyrinth as well as the cortex corticis capillaries empty into them directly.

The vasa efferentia of the juxtamedullary renal corpuscles (Figs. 3 a and 5 a) are the supplying vessels of the renal medulla. At first they may branch to supply the sparse capillary plexus of the outer stripe (which is continuous with the cortical plexuses). When descending further, these efferent arterioles divide to form arterial vasa recta of the vascular bundles. Branches of one vas efferens may contribute to several vascular bundles. Branches may also penetrate into the upper part of the inner stripe independently from a vascular bundle.

Arterial and venous vasa recta together establish the cone-shaped vascular bundles which, in the rabbit, can clearly be pursued deeply into the inner zone. They contain an almost equal number of arterial and venous vasa recta: at the beginning of the inner stripe an average of $\sim$30; at the beginning of the inner zone an average of $\sim$10 arterial and venous vasa recta.

At intervals arterial vasa recta leave the bundles to supply an adjacent capillary plexus, i. e., first the dense-meshed plexus of the inner stripe and then the sparse long-meshed plexus of the inner zone.

The plexuses of the renal medulla are drained by venous vasa recta (Fig. 4), which originate at different levels of the medulla and ascend to the cortico medullary border.

Fig. 4 a-c. Venous medullary vessels filled with microfil. Cross sections *(a)* at the corticomedullary border, *(b)* through the inner stripe of the outer medulla, and *(c)* through the inner medulla. The *broken lines* indicate the approximate borders between cortex *(C)*, outer stripe *(OS)*, inner stripe *(IS)*, and inner medulla *(IM)*. In the inner zone the venous vasa recta are present in small bundles. In the inner stripe they increase and establish a network pattern. The "bridges" *(arrows)* in this network consist of venous vasa recta originating in the inner stripe; the holes *(asterisks)* in this network are occupied by a group of collecting ducts with associated tubules and capillaries. Within the outer stripe the density of the venous vessels is highest; they drain via collecting vessels *(arrow head)* into arcuate veins *(A V)*. *(a)* x $\sim$20; *(b)* x $\sim$20; *(c)* x $\sim$20

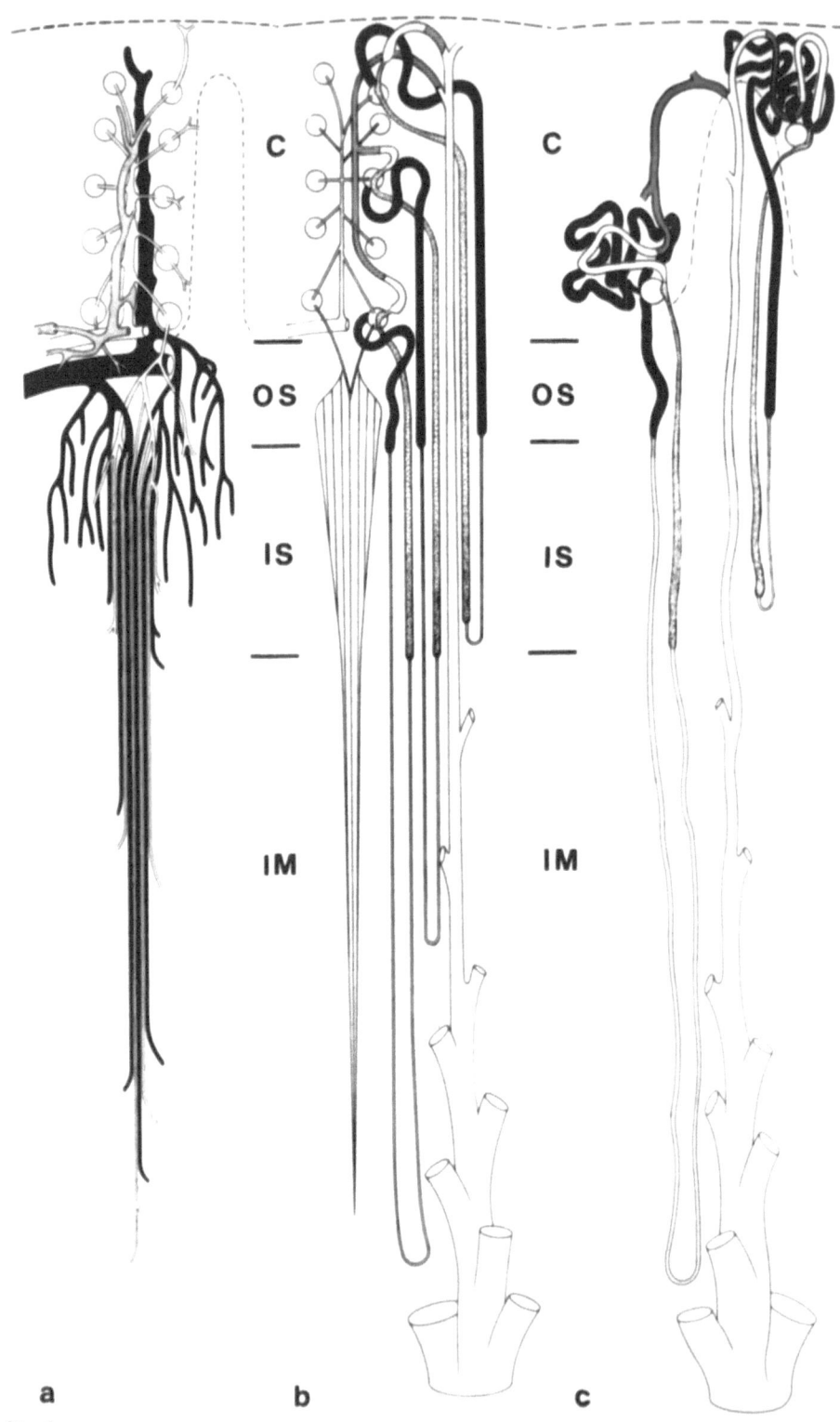

C

OS

IS

IM

a   b   c

Fig. 5

In the inner zone all the venous vasa recta gather within the vascular bundles; (Fig. 4 c) thus, all venous vessels of the inner zone pass the inner stripe within the bundles.

The venous vasa recta draining the inner stripe behave differently (Figs. 3 b and 5 a). Those originating in the lowermost part near the inner zone join the bundles passing through this region. The majority, however, originating in the middle and upper part, ascend directly between the tubules (frequently associated with groups of few descending thin limbs) towards the outer stripe. At the transition of inner and outer stripes the density of the ascending venous vasa recta increases greatly; only in the vicinity of the collecting ducts are no venous vasa recta found. Thus, the ascending venous vasa recta form, as revealed in cross sections (Figs. 4 a and 4 b), a network pattern consisting of the vascular bundles and bridges connecting the bundles. Every hole in this network is occupied by a group of four collecting ducts together with the tubules in their immediate vicinity.

Fig. 5 a–c. a. The renal vessels. Arterial vessels are shown in *white*, venous vessels, *black*, lymph vessels, *gray*. The *broken lines* mark the renal surface and the medullary rays. An arcuate artery gives rise to an interlobular artery from which the vasa afferentia originate at typical angles. The vas efferens of a superficial nephron ascends to the renal surface and the vas efferens of juxtamedullary nephrons splits off into vasa recta. The interlobular vein starts within the cortex corticis. Its basal parts (together with the arcuate vein) accept the medullary venous vessels. The vascular bundle contains all arterial vasa recta and the venous vasa recta from the inner medulla whereas most of the venous vasa recta from the inner stripe ascend independently from the bundle. The lymphatics run together with the arteries. Valves are already present in arcuate lymphatics. b. Basic architectural arrangement of tubules and vessels. The cortical vascular axis of the interlobular vessels is continued in the medulla by vascular bundles. A superficial, a midcortical, and a juxtamedullary nephron is drawn. Proximal tubules are *black*, straight distal tubules are *dotted*, convoluted distal tubules are *white*, connecting tubules are *hatched*, and the collecting ducts are *white*. The loops of Henle are concentrically arranged around the bundles in a manner fully corresponding to their origin in the cortex. The loops of the deepest nephrons are situated nearest to the bundles and those of superficial nephrons, most distant. The distal tubules of juxtamedullary and midcortical nephrons fuse to an arcade, which ascends in the cortical labyrinth, parallel to the interlobular artery. Cytologically an arcade is a connecting tubule. Superficial nephrons drain individually via a connecting tubule into a cortical collecting duct. c. Nephron segmentation as demonstrated in a superficial and a juxtamedullary nephron. The proximal tubule *(black)* of a juxtamedullary nephron is also tortuous in it, straight part (situated exclusively in the outer stripe). The thin descending loop limb *(white)* turns back within the inner medulla and passes over into the straight distal tubule *(dotted)* at the border between inner and outer medulla. Ascending through the outer medulla, the straight distal tubule decreases in diameter. Some short distance beyond the macula densa it passes over into the convoluted part *(white)* which forms only one single coil. By accepting further tributaries, the subsequent connecting tubule *(hatched)* establishes the arcade, which drains into a cortical collecting duct at the border of the cortex corticis. The superficial nephron touches the renal surface with its proximal convolutions *(black)*. The long straight proximal tubule descends within a medullary ray and the outer stripe. The thin limb turns back within the deep part of the inner stripe and passes over into the straight distal tubule *(dotted)*, which has a medullary and a cortical part (the latter lying in the medullary ray). A short distance beyond the macula it passes over into the distal convoluted tubule established by one single coil, which mostly touches the renal surface. In the descending limb of this coil the transition to the connecting tubule *(hatched)* occurs. The connecting tubule returns to the renal surface and the transition to the cortical collecting duct occurs either before or at or after the top of this coil. A cortical collecting duct has short branches by which the connecting tubules are accepted. In the deep cortex and outer medulla collecting ducts do not fuse. In the inner medulla they coalesce and empty in large channels into the renal pelvis. C = cortex; OS = outer stripe; IS = inner stripe; IM = inner medulla

Since in the outer stripe the bundle venous vasa recta also leave the bundles to spread out among the tubules, the total number of medullary draining venous vessels then traverse the outer stripe as wide tortuous channels (Fig. 4 a). Thereby, they behave like capillaries contacting the tubules (the contact to proximal tubules is conspicuously extensive) like capillaries (Figs. 10 a and 12 b). Thus, the blood supply to the outer stripe must be considered predominantly venous. Finally, these venous channels, after joining together, empty into arcuate or into the lowermost parts of interlobular veins. In contrast to other species, almost no venous vasa recta exist which ascend some distance within the medullary rays.

In the context of this paper, it is important to stress that the vascular bundles of the medulla do not join together to form composed bundles in the inner stripe (as, for instance, in the mouse or desert rodents). The number of vascular bundles arising in the outer stripe remains constant in the inner stripe, all reaching the inner zone.

3.3 Lymph Vessels

The lymphatic vessel system in the rabbit kidney (Fig. 5 a) does not principally differ from that in other mammalian species (Kriz and Dieterich, 1970). To understand the functional organization of this drainage system, the lymphatic vessels have to be regarded together with the periarterial loose connective tissue, the extent of which only becomes obvious in quick-frozen specimens (Swann and Norman, 1970; Kriz and Dieterich, 1970). This tissue accompanies the arteries (interlobar, arcuate, interlobular and vasa afferentia) as a circular layer. Its thickness decreases towards the end of the arterial system and vanishes at the entrance of the vas afferens into the glomerulus. In this loose periarterial tissue the lymphatics of the kidney are embedded (Figs. 3 c and 9 c); no other lymphatics exist in the kidney, neither among the tubules of the cortex nor in the medulla.

Generally, the lymphatic capillaries do not start far beyond the periarterial tissue of the interlobular arteries. They leave the kidney as capillarylike channels accompanying the arteries on their route. Occasionally, a lymphatic capillary may be found to accompany a vas afferens (mostly in those cases when the periarterial tissue is somewhat more extensive around a vas afferens) and to enter the vicinity of the vascular pole of a glomerulus (Gorgas, 1978). In addition a lot of renal corpuscles lie in the immediate vicinity of an interlobular artery, i.e., touching the periarterial tissue of such an artery. In these cases a lymphatic capillary can be found in the direct neighborhood of a renal corpuscle. However, it cannot be deduced from these examples that lymphatics are regularly found in the vicinity of a glomerulus. Functionally, however, the difference between these few glomerular vascular poles with an adjacent lymphatic capillary and those without, is negligible, since, in any case, the periarterial tissue (thus the drainage system) starts at the vascular pole of a glomerulus.

A functional interpretation of the lymphatic system has to take into account that the renal cortex is regularly penetrated radially by interlobular arteries and, therefore, also by blind lymphatic channels along these arteries lying in the periarterial loose tissue. This tissue penetrates further into the cortical labyrinth along the vasa afferentia and is continuous to all sides with the intertubular interstitial space of the cortex. Even if this interstitial space among the cortical tubules is sparse (Dieterich and Kriz, 1972; Bulger and Nagle, 1973), each surplus of interstitial fluid produced within the cortical parenchyma may gain access to the periarterial tissue, where it may, on its flowing route along the arteries to the hilus, be gradually reabsorbed by the lymphatics themselves.

3.4 Tubules

3.4.1 General Data

Each kidney of the adult rabbit is composed of ~200,000 nephrons (Rytand, 1938; Bankir and Farman, 1973: 196,200 ± 10,200); the number of nephrons is not correlated to the weight of the kidney (Bankir and Farman, 1973).

According to Peter (1909), the rabbit kidney contains ~60% (58.3%) long-looped nephrons and ~40% (41.7%) short-looped nephrons. We calculated the ratio of short and long loops by counting the loop limbs in relation to the collecting ducts in sections of the outer stripe and of the uppermost inner zone. We found a ratio of 66% long loops to 34% short loops. This somewhat higher figure for the long loops might be due to occasional joinings of collecting ducts occurring between the two counting levels. Thus, the ratio of three long loops to two short loops is regarded as the most approximate value. All short loops have their bend within the innermost third of the inner stripe, while the long loops extend to various levels of the inner zone.

Corresponding to their renal corpuscles, three types of nephrons can be distinguished (Fig. 5): superficial, midcortical, and juxtamedullary nephrons. In addition to this definition of a superficial nephron (originating from a superficial renal corpuscle), another current definition of a superficial nephron says that it touches the renal surface with its convoluted tubules. In so far as proximal coils from midcortical nephrons (the vas efferens of which does not reach the renal surface) can be found on the renal surface, these definitions do not coincide. The subsequent values, which habe been reported by Bankir and Farman (1973), are based on the first definition.

All superficial nephrons (28%) have short loops; other short loops (12%; that is the difference between a total number of about 40% short loops and only 28% superficial nephrons) are supposed to be derived from the uppermost midcortical nephrons. Probably all juxtamedullary nephrons (9%) have long loops; thus, the bulk of long loops (51%) must belong to midcortical nephrons. In summary, the rabbit kidney contains 9% juxtamedullary nephrons (all long looped), 51% long-looped midcortical nephrons, 12% short-looped midcortical nephrons, and 28% superficial nephrons (all short looped).

On the average, six nephrons (6.2 ± 0.3) drain into one cortical collecting duct. The juxtamedullary nephrons and deep midcortical nephrons reach their collecting ducts by "arcades." The superficial nephrons as well as some midcortical nephrons empty individually into the cortical collecting ducts. No joinings of collecting ducts occur within the outer medulla. The collecting ducts unite when entering the inner medulla; the deeper they descend into the inner zone, the shorter the distances become between consecutive joinings. After eight-ten joinings, a few (a constant number probably does not exist) large channels open at the very tip of the papilla or at its side walls into the renal pelvis.

3.4.2 Segmentation

Discussion of the segmentation of the nephron and collecting duct system involves the consideration of different definitions and differences in nomenclature (Valtin, 1977). The difficulty is rooted in the fact that the subdivision of the nephron can be based neither exclusively on histologic criteria nor exclusively on ultrastructural criteria; both are necessary but do not fully coincide. A short introductory summary of the segmentation (Fig. 5 c) might be helpful, although some supporting arguments will not become plausible until the ultrastructural descriptions. The tubular system begins at Bowman's capsule, in most nephrons with a short neck segment. The proximal tubule ist subdivided into a convoluted part and a straight part. The ultrastructural subdivision of the proximal tubule into three segments S1, S2 and S3 (Maunsbach, 1973) overlaps with the histologic subdivisions. The gradual transition from S1 to S2 occurs some distance before the end of the convoluted part and the transition (again gradual) from S2 to S3 occurs within the end portion of the straight part.

The subsequent thin descending limb of Henle's loop is different in long- and short-looped nephrons. In long-looped nephrons it descends through the outer medullary zone into the inner medulla. Therefore it is subdivided into an outer medullary part and an inner medullary part, the latter turning back at various levels of the inner medulla into the thin ascending limb. The thin descending limbs of short loops and the thin ascending limbs of long loops at or near the border between inner and outer zones pass over into the distal tubule.

The term distal tubule is conventionally used to designate that segment of the nephron between the end of the thin limbs in the medulla and the first junction of two tubules in the cortex; the collecting duct system is generally believed to beginn at this junction. Yet this simple definition, based on obvious histologic facts, does not correspond with cytologic criteria. We lay more stress upon the ultrastructural criteria, thereby arriving at the following subdivision, which corresponds to the findings of Peter (1909). The distal tubule is subdivided into a straight part (which is further subdivided into a medullary straight part and a cortical straight part) and a convoluted part. The transition from the cortical straight part to the convoluted part occurs abruptly a short distance beyond the macula densa; thus, the specialized cell plaque of the macula densa is situated within the cortical straight part of the distal tubule.

At a clearly definable border the convoluted part of the distal tubule of each nephron passes over into the connecting tubule, regarded as the beginning of the collecting duct system. This tubular segment exhibits the same ultrastructural characteristics in all nephrons. As regards histologic criteria, it is developed differently in juxtamedullary and deep midcortical nephrons as compared to more superficially situated nephrons. In juxtamedullary and deep midcortical nephrons the connecting tubules come together to form a joined connecting tubule, histologically known as an arcade. These arcades ascend within the cortical labyrinth, looping at the level of the most superficial layer of glomeruli toward the medullary ray to pass over into a cortical collecting duct. The connecting tubules of superficial and upper midcortical nephrons do not join but pass over individually into a cortical collecting duct.

Based on considerable but always gradual alterations in the epithelial lining, the collecting ducts are subdivided with respect to localization into cortical collecting ducts, outer medullary collecting ducts, and inner medullary collecting ducts. The latter join together successively and finally open into the renal pelvis at the papillary

14

tip. The term inner medullary collecting duct also includes the distal most segments, also called papillary ducts (ducts of Bellini).

3.4.3 Nephrons

(In cooperation with H. Koepsell, Max Planck Institute for Biophysics, Frankfurt/ Main, FRG)

Superficial nephrons (Figs. 6 a and 8). The renal corpuscles of superficial nephrons lie beneath the cortex corticis, at least 400-μm distant from the renal surface. The convoluted portions (proximal and distal) and the connecting tubules of these nephrons are generally situated above the renal corpuscles and form the cortex corticis.

The straight portions (proximal and distal) of these nephrons together with the top portions of the cortical collecting ducts constitute the apical parts of the medullary rays. In most cases, the descending and ascending straight portions of the same nephron run side by side over the total length of the medullary ray (Fig. 6 a_2). Moreover, both limbs either lie side by side or at least near to the corresponding collecting duct. Generally, the straight distal tubule is more often placed in the direct neighborhood of its corresponding collecting duct than is the straight proximal tubule. The course of these nephron portions (straight proximal and distal) is actually straight over their whole length, i. e., within the medullary rays and within the outer stripe.

In the outer stripe (Fig. 6 a_3) the corresponding proximal and distal straight portions keep their histotopographic relationships, neighboring each other and near their own collecting duct. In contrast to the juxtamedullary nephrons, they are distant from the vascular bundles. Some distance before the border to the inner stripe and somewhat earlier than in juxtamedullary nephrons, the straight proximal tubules pass over into the thin descending limbs.

In the inner stripe (Fig. 6 a_{4-5}) the descending thin limbs assume a fully comparable position to that of the straight proximal tubules of the superficial nephrons in the outer stripe: generally neighboring their counterparts, both corresponding descending and ascending limbs are situated near their collecting duct and distant from the vascular bundles.

The bends of these loops (Fig. 6 a_6) have all been found in the innermost part of the inner stripe. The transition from the thin descending limb to the thick-wall, straight distal tubule may occur (without any preference) either shortly before the bend, at the bend itself, or a short distance after the bend (Fig. 6 a_5). Since both loop limbs run side by side, the bend between both is sharp.

Midcortical nephrons (Figs. 6 b and 8). In the rabbit kidney midcortical nephrons may have either (probably the more superficial ones) short loops or (the majority) long loops.

The course of a short-looped midcortical nephron is similar to that of a superficial nephron. The main difference is found in the convoluted portions. In a midcortical nephron (as evidenced in our injections), the convoluted portions do not ascend into the cortex corticis, but are mainly localized in the neighborhood of their glomerulus. However, microdissection (Bankir et al., 1975) has clearly shown that there are midcortical nephrons whose proximal tubules with one or two coils may ascend through the cortex corticis and touch the renal surface. The loop of Henle, within the

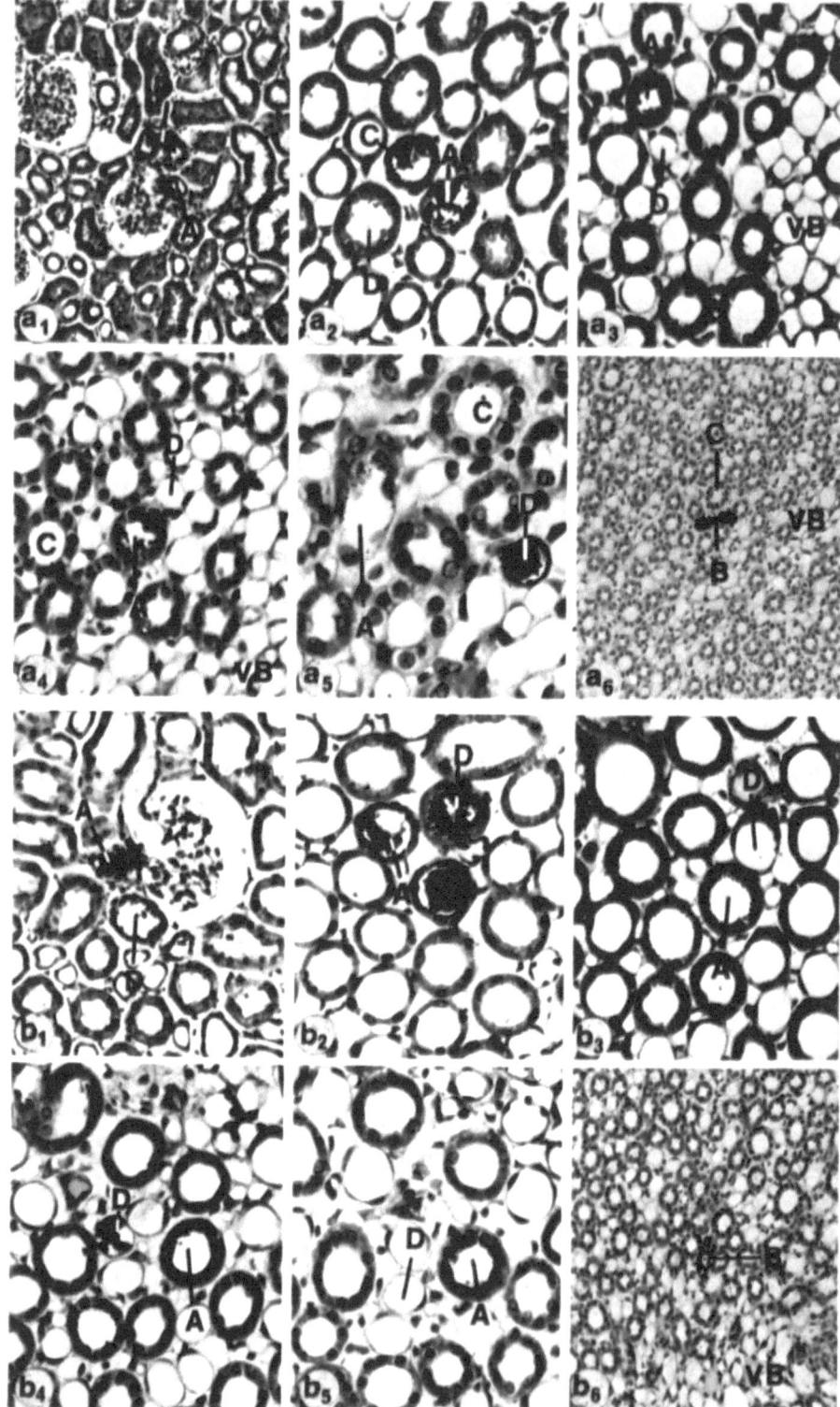

Fig. 6

medullary ray (Fig. 6 b$_2$) and within the outer stripe (Fig. 6 b$_3$) as well as the inner stripe (Fig. 6 b$_{4.6}$), exhibits histotopographic relationships comparable to the loop of a superficial nephron.

It may be assumed that the straight proximal and straight distal tubules of mid-cortical nephrons with long loops (we did not manage to inject a nephron that we could clearly call a midcortical nephron with a long loop) run through the lower half of the medullary rays. In the outer and inner stripe they probably occupy a position in between the juxtamedullary loop limbs (neighboring a vascular bundle) and the superficial loop limbs (neighboring a collecting duct). Since the distance between a vascular bundle and the concerned collecting ducts ist not too far and variable, a good deal of intermingling of the loops within the inner stripe should be expected.

In the inner zone differences in the histotopographic relationships between the loop limbs of a midcortical and a juxtamedullary nephron are not imaginable. Differences may exist in respect to the actual length of the loop; it seems reasonable to suppose that the majority of the longest long loops extending into the tip of the papilla are derived from the deepest juxtamedullary nephrons, whereas the majority of the shorter long loops are derived from midcortical nephrons; however, exceptions have been found.

Juxtamedullary nephrons (Figs. 7 a, 7 b and 8). The true juxtamedullary nephrons (by definition only those with efferent arterioles contributing to the vascular bundles) amount to only ~9% of all nephrons; their renal corpuscles are usually situated in the innermost cortex. The convoluted portions (Figs. 7 a$_1$ and 7 b$_1$) of these nephrons establish the deepest parts of the cortical labyrinth. They emerge as relatively small bars of transversely arranged tubules separating the basal parts of the medullary rays (Fig. 7 a$_1$).

The straight tubular portions of juxtamedullary nephrons are not found within the medullary rays. The "straight" portions of the proximal tubules (Figs. 7 a$_{2-3}$, 7 b$_{2-3}$) directly penetrate into the outer stripe, traversing it in a somewhat tortuous manner in the immediate vicinity of a vascular bundle. The course of the straight portions of the distal tubule is, in contrast to their proximal counterparts, well characterized by the term straight. The straight distal tubules of juxtamedullary nephrons are also situated near a vascular bundle; they reach their renal corpuscle directly without traversing the medullary rays.

In the inner stripe (Figs. 7 a$_{4.5}$ and 7 b$_{4.5}$) both loop limbs, the thin descending limb, as well as the thick ascending limb (straight portion of the distal tubule) retain their positions in the vicinity of a vascular bundle (Fig. 8 e-m). Both corres-

---

Fig. 6 a and b. Short looped nephrons shown in cross sections after microfil injection. D = proximal tubule resp. descending limb; A = distal tubule resp. ascending limb; B = bend; C – corresponding collecting duct; VB = vascular bundle. a. Superficial nephron: (a$_1$) renal corpuscle and convoluted portions; (a$_2$) medullary ray; (a$_3$) border of outer and inner stripe; transition of the proximal epithelium to the thin descending limb epithelium; (a$_{4-6}$) inner stripe; (a$_5$) transition of the thin to the thick ascending limb epithelium; (a$_6$) bend, lined by thin epithelium. b. Midcortical nephron: (b$_1$) portions of the proximal and distal tubule at the level of the macula densa; (b$_2$) medullary ray; (b$_3$) border of outer and inner stripe; transition of the proximal epithelium to the thin descending limb epithelium; (b$_{4-6}$) inner stripe; (b$_6$) bend. (a$_1$, a$_2$, a$_6$) x ~130; (a$_5$) x ~250; (b$_1$) x ~500; all others, x ~330

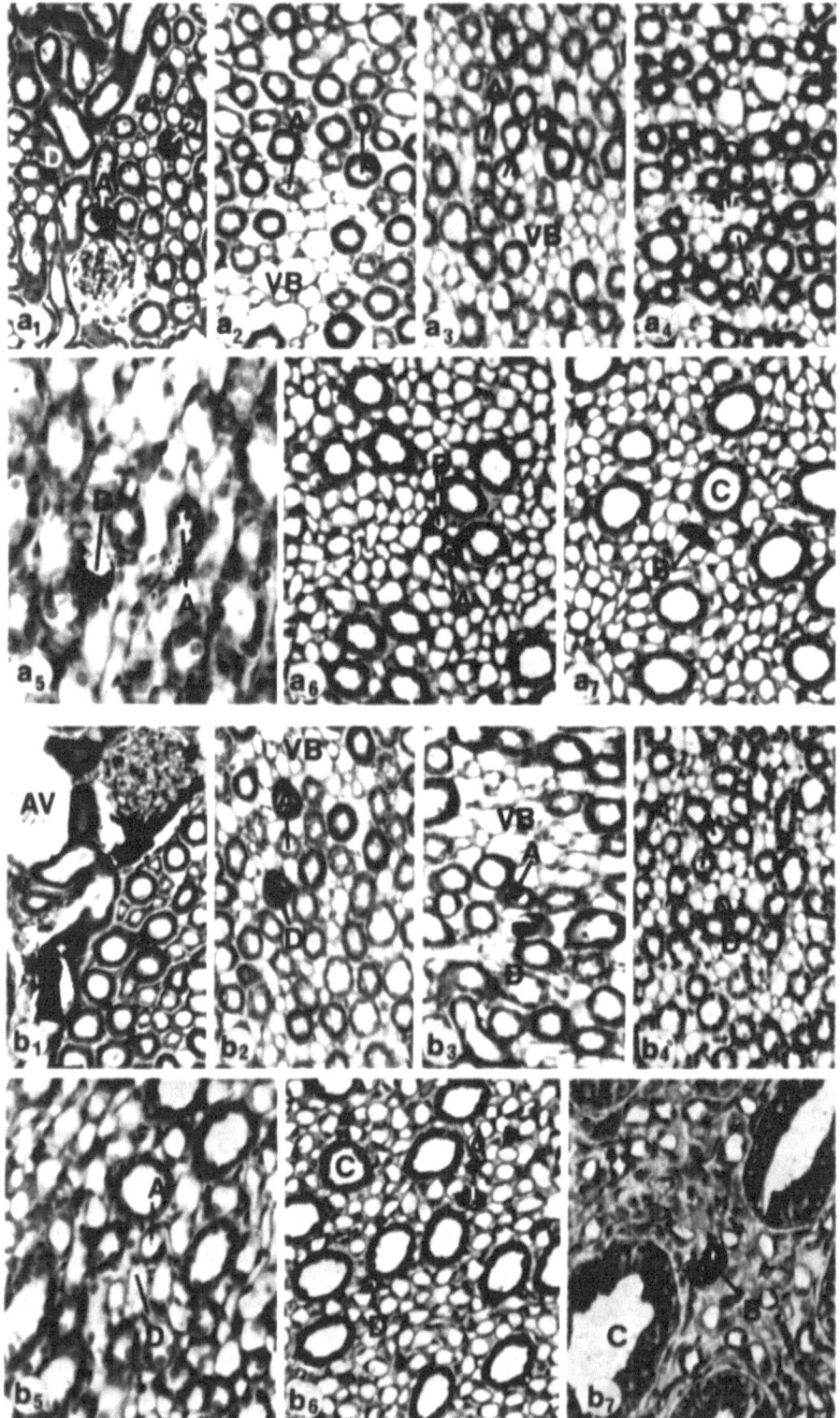

Fig. 7

ponding loop limbs mainly run side by side. In the inner medullary zone (Figs. 7 a 6-7, 7 b 6-7 and 8 q) characteristic histotopographic relationships are obvious neither for the descending nor for the ascending loop limbs. Both corresponding thin limbs run parallel and neighbor each other; both limbs may be situated either next to a collecting duct or distant from it. It may be assumed that most of the juxtamedullary nephrons reach the papilla (Fig. 7 b 7). The bend may be sharp or broad.

Contacts of the distal tubule to arterial vessels. The contact point between the macula densa of the distal tubule and the vascular pole of its corresponding glomerulus is clearly established in each nephron (Fig. 3 c); the macula densa generally broadly faces the Goormaghtigh cell field (Fig. 32). As we did not systematically investigate the question as to whether the macula, in addition to its contact to the Goormaghtigh cells, is more related to the vas afferens than to the vas efferens, no substantial data can be contributed.

In addition to the contact of the macula densa to the arterial pole of the corresponding glomerulus, contacts between the distal tubule and the vas afferens and vas efferens have been described in the rat (Barajas and Latta, 1963; Faarup, 1965, 1971; Gorgas, 1978). In the rabbit we also found conspicuous histotopographic relationships of the distal tubule to arterial vessels. We did not investigate them systematically, thus, they will only be briefly mentioned.

The straight part of the distal tubule touches the efferent arteriole either with its premacula portion or with its short postmacula segment both closely and over long distances (Fig. 9 d). In juxtamedullary nephrons, the premacula portion probably contacts the efferent arteriole and in midcortical and superficial nephrons, the postmacula segment. Since these contacts occur so often, we believe them essential to each nephron. No other tubular segments were found as regularly and as near in the vicinity of the efferent arteriole.

Fig. 7 a and b. Long looped nephrons shown in cross sections after microfil injection. D = descending limb; A = ascending limb; B = bend; C = collecting duct; VB = vascular bundle; AV = arcuate vein. *a.* Nephron with its loop bend in the high inner zone: (a_1) proximal and distal tubule, cut at the level of the macula densa; (a_2) early outer stripe; (a_3) border between outer and inner stripe with the transition of the proximal tubule epithelium to the thin descending limb epithelium; (a_4) inner stripe; (a_5) border between inner stripe and inner zone, transition from the thin to the thick ascending limb; (a_6) inner zone; (a_7) inner zone, bend. *b.* Nephron with its loop bend in the tip of the papilla: (b_1) renal corpuscle and proximal convolutions; (b_2) early outer stripe; (b_3) border between outer and inner stripe with the transition of the proximal tubule epithelium to the thin descending limb epithelium; (b_4) inner stripe; (b_5) border between inner stripe and inner zone, transition of the thin to the thick ascending limb epithelium; (b_6) high inner zone; (b_7) deep inner zone, bend. (a_1, b_1) x ~100; (a_5) x ~330; (b_5) x ~260; all others, x ~130

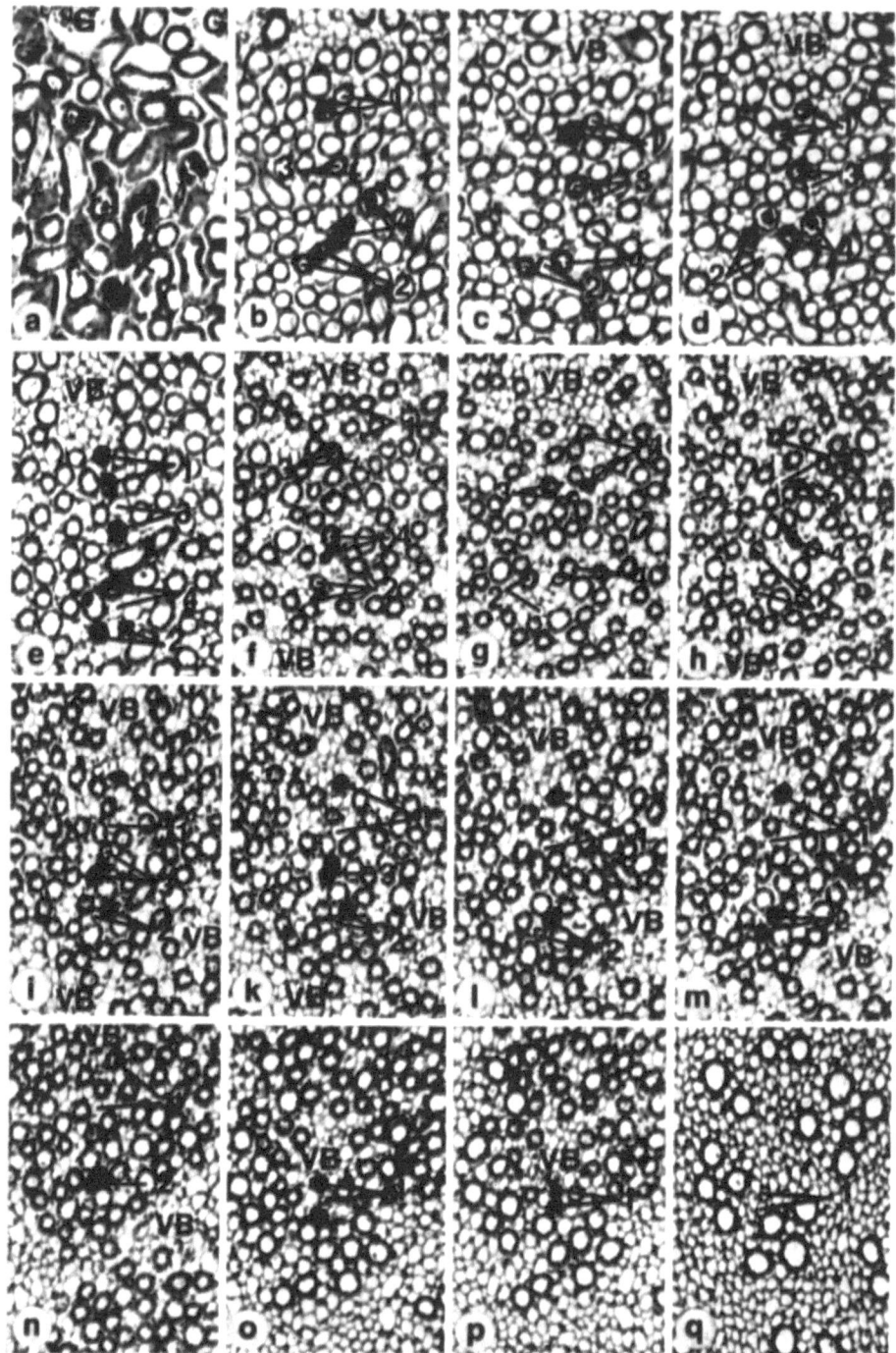

Fig. 8

3.4.4 Collecting Duct System

The superficial nephrons (and the upper midcortical nephrons) drain individually via a connecting tubule into a cortical collecting duct (Fig. 5 b and 36 a-c). In superficial nephrons the short convoluted part forms a single coil, which touches the renal surface. After looping back, it passes over into the connecting tubule, which may return with a single coil to the surface. However, the transition from the connecting tubule to the cortical collecting duct (defined by ultrastructural criteria; see 4.4) may occur either before or behind the actual bend of this latter coil. Accordingly, in addition to proximal and distal tubules, connecting tubules and cortical collecting ducts may be found on the renal surface (Fig. 5 c).

The juxtamedullary nephrons and deep midcortical nephrons with their connecting tubules join to form the arcades (Figs. 5 b-c and 36 b). The latter ascend in the cortical labyrinth, parallel and next to the interlobular vessels, separated from the interlobular arteries only by the periarterial loose connective tissue sheet (Fig. 9 b and c). At the level of the uppermost layer of renal corpuscles, they turn laterally to loop into a medullary ray where they join a cortical collecting duct (Figs. 9 b, 36 d and f).

Each medullary ray contains four collecting ducts (Figs. 9 a and 12 a). At their origin they extend short branches to pick up the superficial nephrons (Fig. 36 c); somewhat deeper, again with short branches, they pick up the arcades (Figs. 5 c, 9 b and 36 d–f). On their descending route through the medullary rays additional tributaries from individual midcortical nephrons may be accepted. There are no joinings of collecting ducts within the medullary rays and the outer medulla. Thus, the groups of four collecting ducts corresponding to the original medullary ray status are maintained until the beginning of the inner zone (Figs. 10, 11 a, 12 and 13). Here they begin joining; the first two joinings belong to the same group. The deeper they descend into the inner zone, the shorter the distances are between the consecutive joinings.

3.5 Architecture

The kidney is described as composed of individual architectural units. Usually the collecting ducts of one medullary ray are considered the center of a renal lobule; all nephrons draining into these ducts establish the lobule (among others Oliver, 1968). According to another concept (v. Möllendorff, 1930; Smith, 1951; Kriz, 1967), an "interlobular" artery of the cortex establishes the central axis of a lobule unit. This may be called a vascular lobule; all nephrons originating from this artery together

---

Fig. 8 a-q. Microfil injection of one nephron *(1)* with a long loop and three nephrons *(2, 3, 4)* with short loops. A superficial nephron was injected from the kidney surface. The others were filled by retrograde flow. *VB* = vascular bundle. *a.* Cortex; proximal and distal convolutions; *b.* outer stripe; *c.* and *d.* border of outer and inner stripe; *c.* transitions of the proximal epithelium to the thin descending limb epithelium of *2* and *3; e–h.* inner stripe; *h.* bend of nephron *4, λ.* bend of nephron *3; n.* bend of nephron *2,* extending almost to the border of inner stripe and inner zone; *o.* and *p.* border between inner stripe (in the *upper half*) and inner zone (in the *lower half* of the photo); *q.* inner zone, thin descending and thin ascending loop limbs of nephron *1.* Note that the limbs of the long loop both lie in the vicinity of the vascular bundle, whereas the short loop limbs mostly lie in the vicinity of collecting ducts. *a–q* x ~130

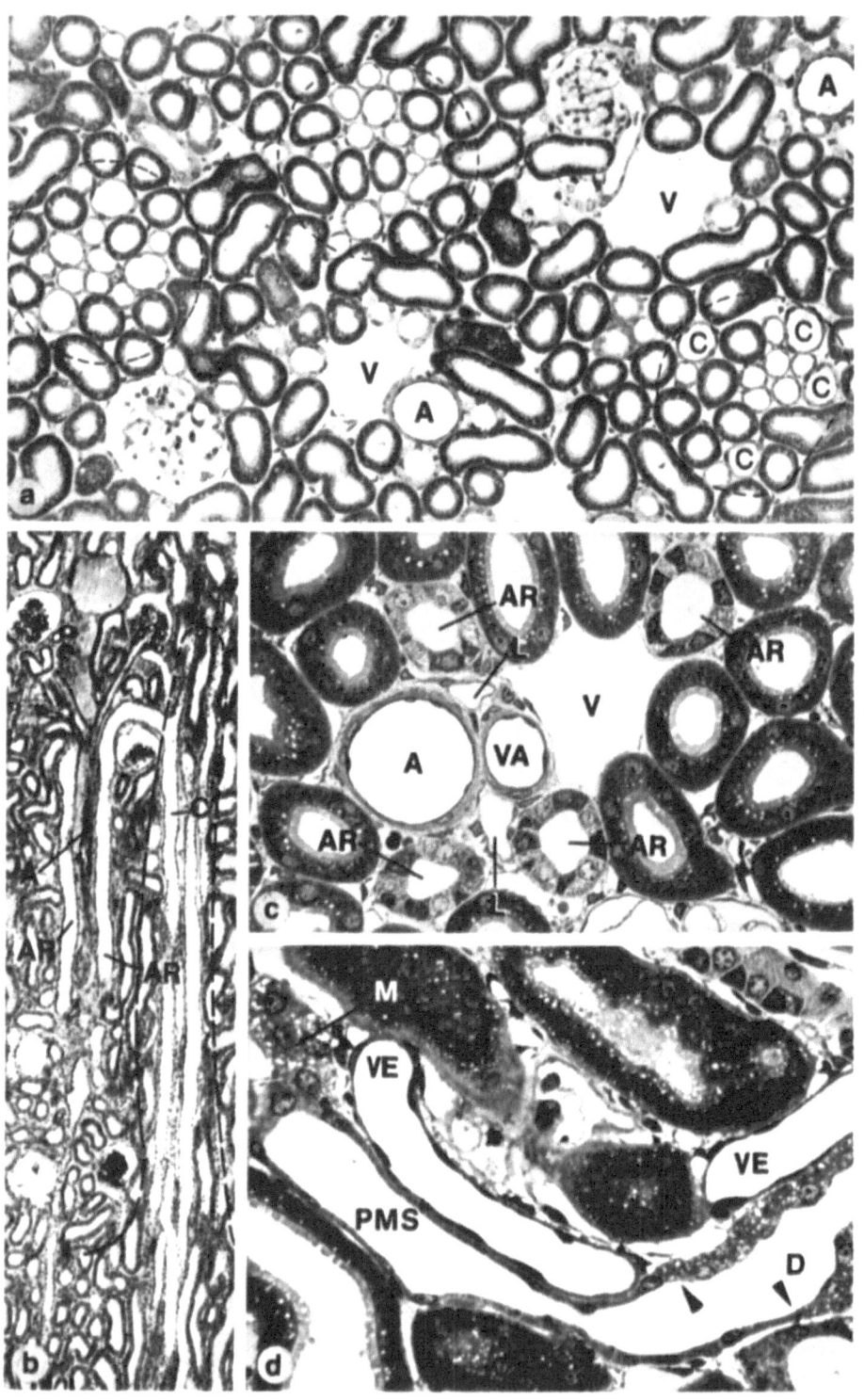

Fig. 9

with their related collecting ducts constitute the lobule. Certainly both lobular concepts are suitable for describing the architecture of the kidney.

There is not much sense in arguing which is the actual lobule, because neither really exists nor can they be delimited in histologic sections. Therefore, it is much more reasonable to pose the question in the following way: which concept is more suitable for explaining the architecture (and functional morphology) of the kidney, and which one is more useful for comparative anatomic descriptions? From these points of view, the adoption of the vascular lobule-unit-concept clearly offers many advantages.

An "interlobular" (preferably a "lobular") artery constitutes the axis of a vascular lobule. In its surroundings the nephrons with their loops of Henle are arranged like pendant festoons with the loops of juxtamedullary nephrons close to the axis and those of the superficial nephrons most distant from it. The collecting ducts are not precisely situated at the very border of such a lobule, but are incorporated within the corresponding loops of Henle. On the whole, the loops of superficial nephrons lie more peripherally and those of midcortical and juxtamedullary nephrons, more centrally (Fig. 5 b).

The vascular bundles are considered to continue the lobule axis into the medulla. However, this is not always correct, since branches of a juxtamedullary vas efferens may well contribute to different bundles. Moreover, in certain species vascular bundles may fuse (e.g., giant bundles in Psammomys), thereby combining several cortical lobules in a single medullary unit. Although the weak points of the model become evident at this point, the conceptual value of the vascular lobule unit cannot be questioned.

3.5.1 Cortex

The architectural pattern of the cortex is best understood by means of cross sections through its middle part (Fig. 9 a, c). Its essential features are represented schematically in Fig. 12 a. The "interlobular" artery, referred to as axis of the lobule, is accompanied by an interlobular vein and a lymphatic vessel, which is inserted into the periarterial loose connective tissue sheet. Immediately around this vascular axis (separated from the artery only by the periarterial connective tissue sheet), the arcades

---

Fig. 9 a – d. Cortex. *a.* Semithin cross section through a middle level of the cortex containing three medullary rays (demarcated by a *broken line*), each of which contains four cortical collecting ducts *(C)*. In the cortical labyrinth an interlobular artery *(A)* and an interlobular vein *(V;* which may be situated somewhat distant from the ortery) establish the vascular axis. x ~135. *b.* Longitudinal section (paraffin embedded material) through the cortex; the medullary ray is demarcated by a *broken line*. Within the cortical labyrinth two arcades are parallel to the interlobular artery *(A);* one arcade turns laterally (just beneath the most superficial layer of glomeruli) to join a cortical duct *(C).* x ~100. *c.* Semithin cross section demonstrating a vascular axis with an interlobular artery *(A),* which has given rise to a vas afferens *(VA); V* = interlobular vein. Two lymphatics *(L)* lie in the periarterial connective tissue. Around the artery three arcades *(AR)* are arranged, a fourth arcade touches the vein. x ~340. *d.* Semithin longitudinal section demonstrating the close relationships of a vas efferens *(VE)* to the postmacula segment *(PMS)* of the straight distal tubule as well as to the convoluted distal tubule *(D). M* = Macula densa; *arrow heads* indicate the transition from the postmacula segment to the convoluted distal tubule. x ~380

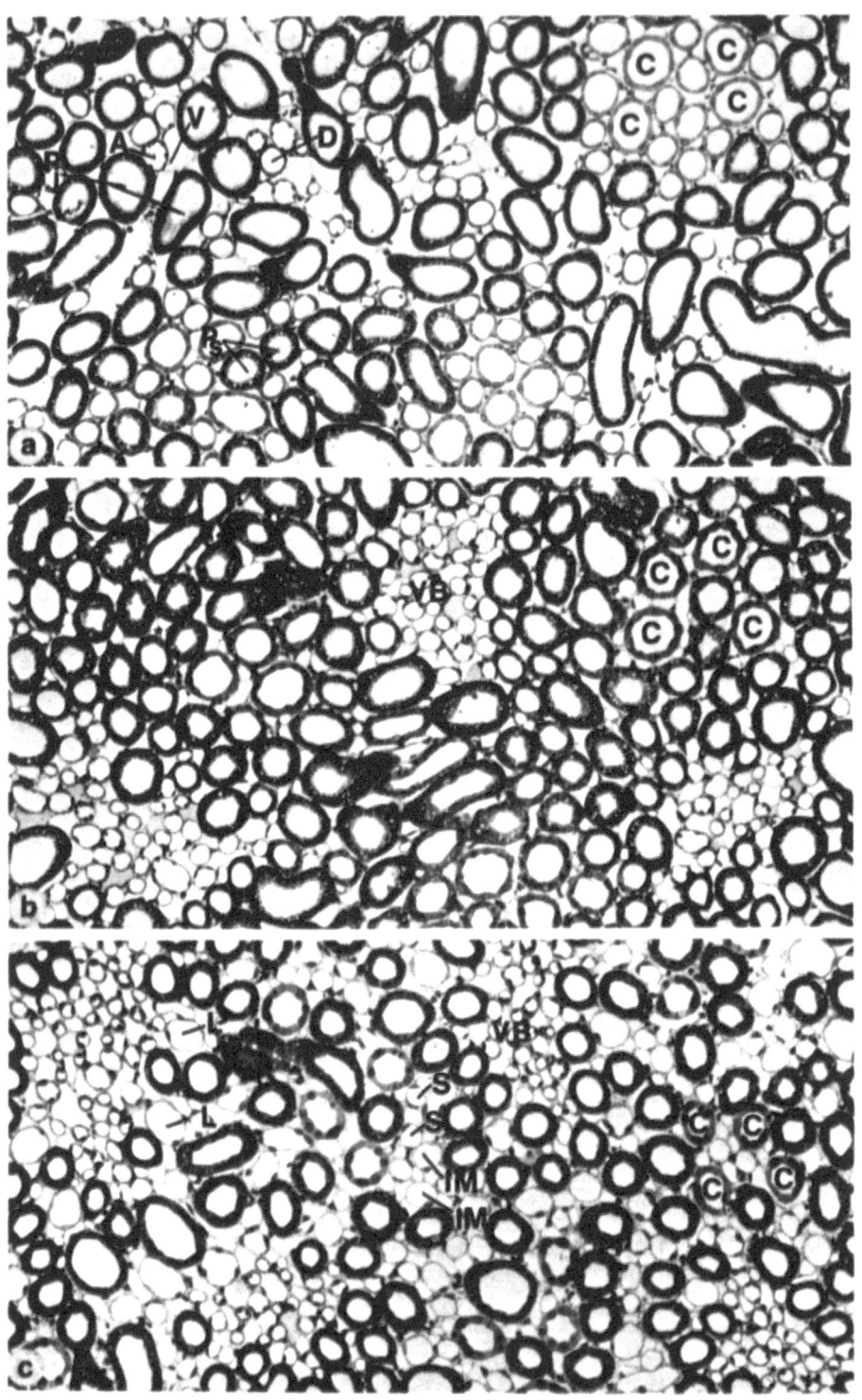

Fig. 10

are constantly localized (Fig. 9 c). At least one and more frequently two, three, or up to four arcades may be found parallel to an interlobular artery (Fig. 9 b): an arcade in another histotopographic position was not found. The vasa afferentia leave the lobule axis, wrapped in some periarterial loose connective tissue, to find their glomeruli. On the whole, the vasa efferentia continue the course of the vasa afferentia, finally branching up into capillaries usually near the border of the labyrinth. The vascular pole of a glomerulus is traversed by the straight distal tubule, at this site forming the macula densa and then passing over some short distance beyond the vascular pole into the convoluted distal tubule.

The medullary rays comprise the straight proximal and distal tubules together with the cortical collecting ducts. Usually, a medullary ray contains four collecting ducts, situated neither in the very center nor at the periphery of a medullary ray (Figs. 9 a and 12 a). They are pushed away from the periphery by straight portions of midcortical nephrons; on the other hand, it can easily be supposed that the portions of superficial nephrons form the center of a medullary ray. In the vascular lobule the collecting ducts of one medullary ray belong to different lobules; an exact delimitation of the lobule border is not possible.

The cortex corticis in the rabbit is a well-developed layer ~ 0.5-mm thick and covering the total surface of the kidney. It contains the convoluted portions of the superficial nephrons (as well as some coils of tubules of deeper i. e., upper midcortical nephrons), thus forming part of the cortical labyrinth. Out of 100 tubules touching the renal surface − and thus being accessible for micropuncture − 90 are proximal convoluted tubules, 5 are distal convoluted tubules, and 5 are connecting tubules or cortical collecting ducts.

3.5.2 Medulla

The architecture of the renal medulla is best understood by cross-sectioning on subsequent levels down to the papilla (Figs. 10 and 11). In addition, the essential features of the architectural pattern are demonstrated in schemes (Figs. 12, 13 and 14).

Outer stripe (Figs. 10, 12 and 13). A cross section through the upper part of the outer stripe near to the cortex shows the developing vascular bundles, which continue along a cortical lobule axis. This cross section consists of a few arterial and venous vasa recta, which are still intermingled with the "straight" proximal tubules of

---

Fig. 10. Outer medulla. Semithin cross sections through *(a)* the upper part of the outer stripe, *(b)* the lower part of the outer stripe, and *(c)* a middle level of the inner stripe. The original pattern of the cortex (compare to Fig. 9 a) is maintained: groups of four collecting ducts *(C)* mark the continuations of the medullary rays. In *(a)* the vascular bundles are not yet established; the vasa recta (arterial = *A;* venous = *V)* are intermingled with straight proximal *(P)* and distal *(D)* tubules. The "straight" proximal tubules of juxtamedullary nephrons *(Pj)* are conspicuously thicker in outer diameter than those of more superficial nephrons *(Ps)*. At the end of the outer medulla *(b)* the vascular bundles *(VB)* are fully established. Also in the inner stripe *(c)* a vascular bundle contains only arterial and venous vasa recta. The heterogeneity among the thin loop limbs is obvious. Large cross-sectional profiles *(L;* supposedly derived from juxtamedullary nephrons) are predominantly situated near the vascular bundles; whereas intermediate *(IM)* and small *(S)* profiles tend to lie distant from the bundles. *(a−c)* x ~ 135

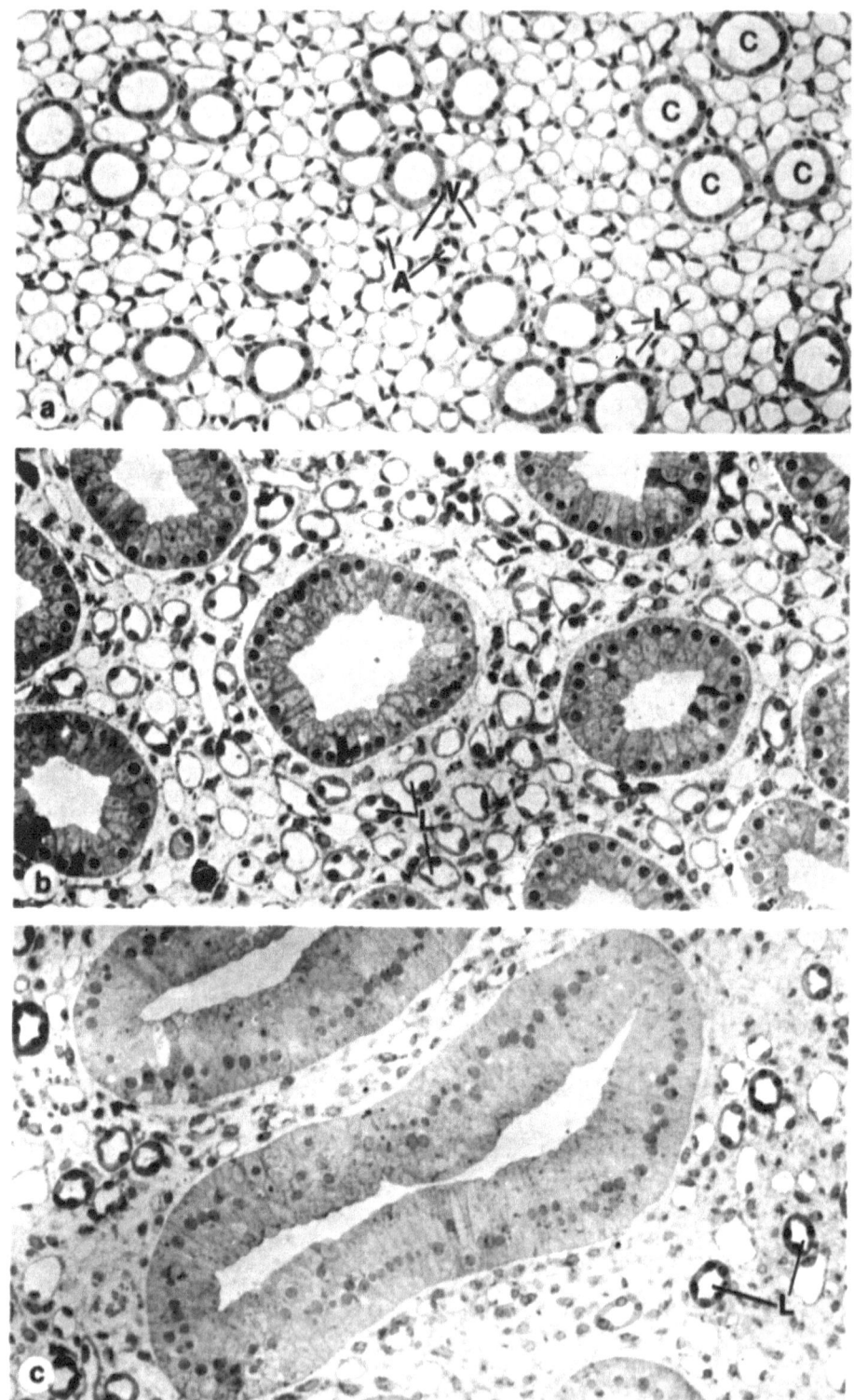

Fig. 11

juxtamedullary nephrons. Obviously the proximal tubule extensively contacts venous vasa recta. The bulk of straight distal tubules of juxtamedullary nephrons lies somewhat more peripherally. The tubules continuing the medullary rays are arranged around the center of vasa recta and juxtamedullary straight nephron portions; their pattern has not changed. However, venous vasa (those coming up directly from the inner stripe) have partially replaced the capillaries.

A comparison at this site of the straight portions of juxtamedullary nephrons with those of midcortical and superficial nephrons shows regular differences in diameter and cellular thickness. The straight proximal tubules of juxtamedullary nephrons are much larger than those of midcortical and superficial nephrons ($\sim$65 μm compared to $\sim$ 50 μm in diameter); in contrast, the straight distal tubules of juxtamedullary nephrons, at this site, are smaller in diameter and cellular thickness ($\sim$25 μm compared to 30 μm for the outer diameter; $\sim$4 μm compared to 6 μm for the cellular thickness).

At a deeper level of the outer stripe near the transition to the inner stripe (Figs. 10 b and 13 a), the vascular bundles have reached their final dimensions. They are composed of an almost equal number of arterial and venous vasa recta (about 30 of both types) arranged in an alternating pattern. The loop limbs (straight proximal and distal tubules) of juxtamedullary nephrons are situated just around the bundles, followed by the limbs of midcortical and superficial nephrons, and by the collecting ducts. The proximal tubules of midcortical and superficial nephrons pass over into the thin descending loop limbs somewhat earlier than those of juxtamedullary nephrons; therefore, in the lower part of the outer stripe, thin descending limbs or at least proximal thin-limb transitions of midcortical and superficial nephrons are found. The corresponding somewhat deeper transitions of the juxtamedullary nephrons constitute the border between outer and inner stripe.

The differences in diameter and cellular thickness between straight tubules (proximal and distal) of juxtamedullary nephrons on one side and those of midcortical and superficial nephrons on the other side have become even more evident than in the upper part of the outer stripe.

Inner stripe (Figs. 10 c, 13 b and 21). The architectural pattern of the inner stripe remains fundamentally unchanged as compared to that of the outer stripe. The vascular bundles do not decrease in number; they represent primary bundles (i.e., no fusions) regularly distributed over the entire cross-sectional area. They contain an approximately equal number of arterial and venous vasa recta.

The loops of Henle are arranged around the bundles in essentially the same pattern as in the outer stripe. Although they increasingly intermingle, the loops of

---

Fig. 11. Inner medulla. Semithin cross sections through *(a)* the upper part, *(b)* the middle part of the inner zone, and *(c)* the tip of the papilla. In *(a)* the collecting ducts *(C)* are still arranged in groups of four. The vascular bundles cannot be clearly delimited. Vasa recta *(A; V)*, thin loop limbs *(L)*, and capillaries together occupy the space between the collecting duct groups. In *(b)* the collecting ducts belonging to one group have fused; the collecting ducts have tremendously increased in diameter and cellular thickness. Vasa recta, capillaries, and thin loop limbs are randomly distributed between the collecting ducts and the interstitial space has increased. At the tip of the papilla *(c)* the collecting ducts have reached their ultimate size. Within the abundant interstitial space vessels and loop limbs *(L)* are sparse. Loop limbs have also increased in diameter and cellular thickness. *(a–c)* x ~174

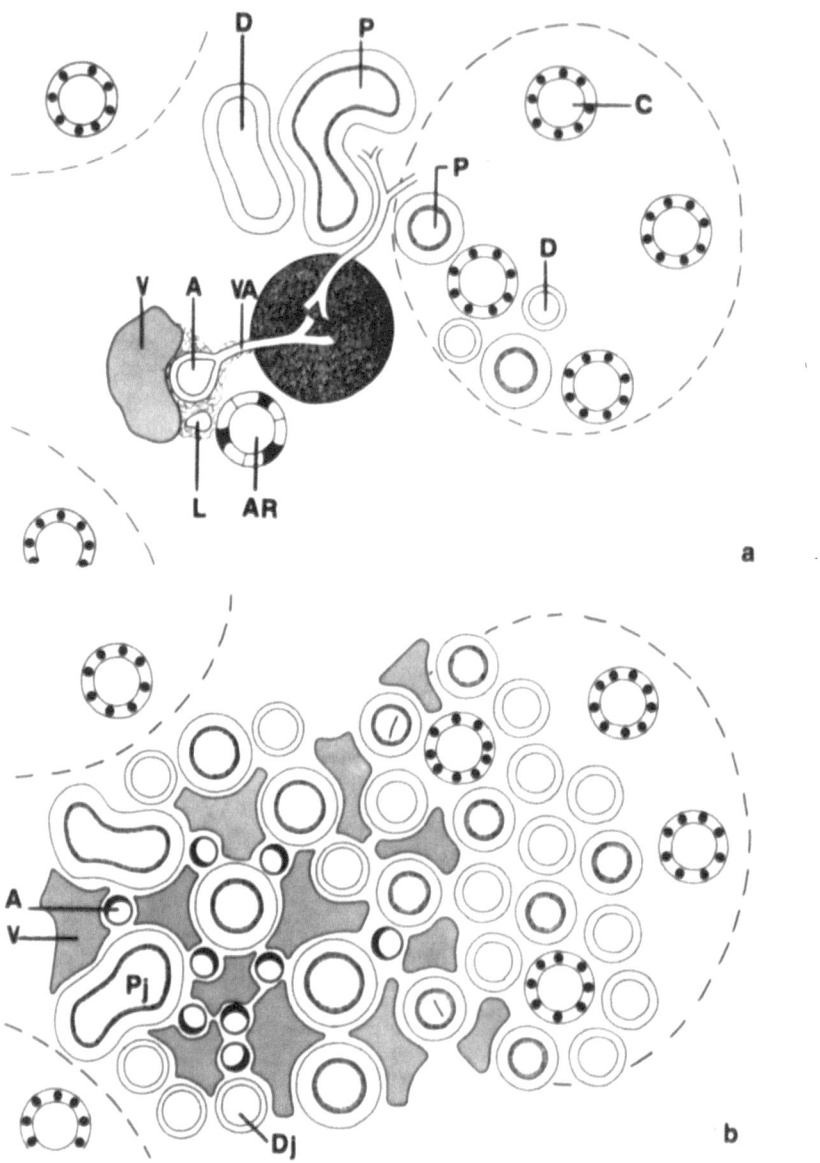

Figs. 12–14. The architectural pattern by a sequence of schematically drawn cross sections through six different levels from the cortex to the papilla. The cortex (12 a) consists of the cortical labyrinth and the medullary rays (demarcated by a *broken line*). In the labyrinth, an interlobular artery *(A)* and vein *(V)* establish the vascular axis. The artery is wrapped by some loose connective tissue containing a lymphatic *(L);* the connective tissue generally continues along the vas afferens *(VA)* to the glomerulus. The vas efferens splits off into capillaries at the border to the medullary ray. Within the labyrinth the convoluted proximal *(P)* and distal *(D)* tubules are situated. In addition, the arcades *(AR)* ascend parallel to the interlobular vessels. The medullary ray regularly contains four collecting ducts *(C)* and the straight parts of proximal *(P)* and distal *(D)* tubules.

In the upper part of the outer stripe *(12 b),* the vascular bundles are not yet fully established. Arterial *(A)* and venous *(V)* vasa recta are intermingled with straight proximal and distal tubules. The straight proximal *(Pj)* and distal *(Dj)* tubules of juxtamedullary nephrons are situated near the

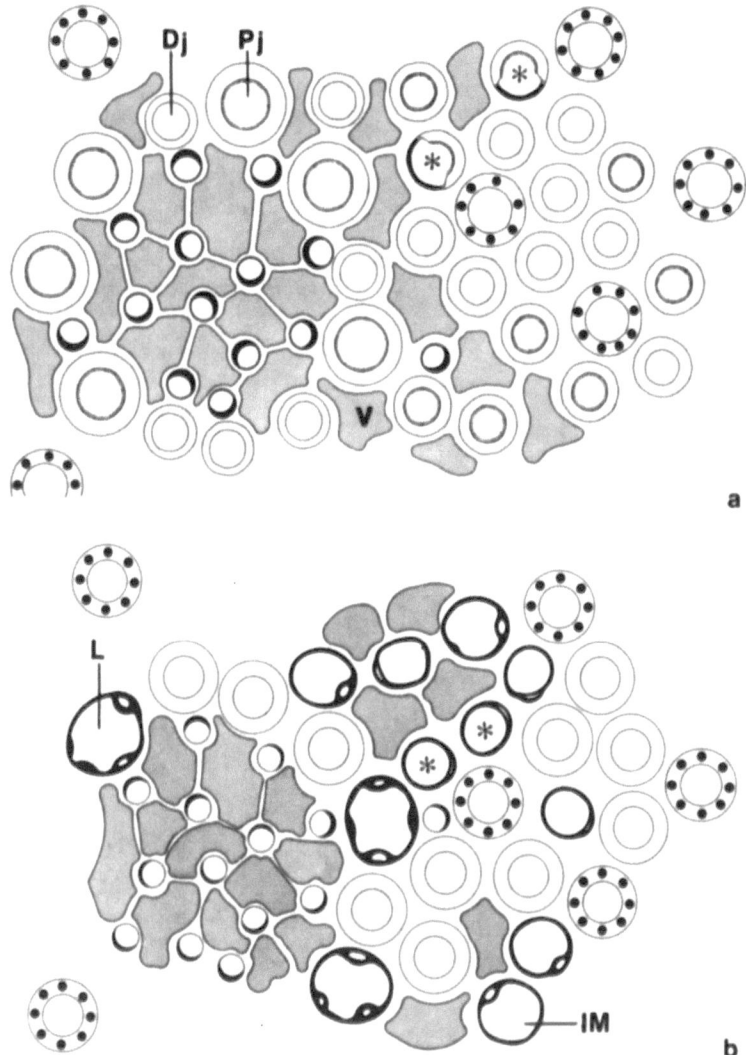

Fig. 13

Fig. 12–14 contd.

developing center of the vascular bundles. The straight proximal and distal tubules of all more superficially originating nephrons are continuations of the medullary rays and are arranged around the collecting ducts. The straight proximal tubules of juxtamedullary nephrons clearly surpass those of midcortical and superficial nephrons in diameter.

In the lower part of the outer stripe *(13 a)* the vascular bundles are fully developed. Those venous vasa recta *(V)* lying outside the bundles ascend directly from the inner stripe. The straight proximal *(Pj)* and distal *(Dj)* tubules of juxtamedullary nephrons lie closest to the vascular bundles, whereas those of midcortical and superficial nephrons are arranged around the collecting ducts; note the difference in outer diameter and cellular thickness between the two groups. Straight proximal tubules of superficial (or midcortical) nephrons pass over somewhat earlier into the thin descending limb of Henle's loop *(asterisk)* than those of juxtamedullary nephrons.

In the inner stripe *(13 b)* the overall arrangement (juxtamedullary nephrons near to the bundles, superficial nephrons most distant from them) is maintained; the collecting ducts are still gathered in groups of four. The descending thin limbs are heterogeneous with respect to the outer diameter

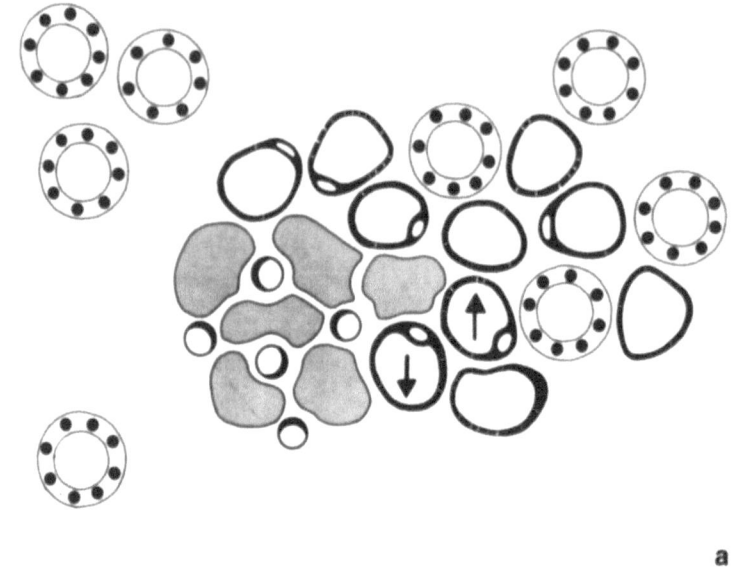

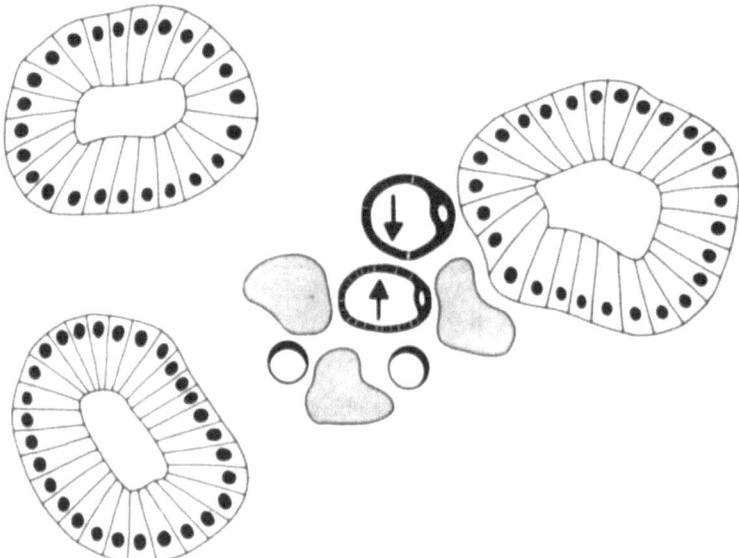

Fig. 14   b

Fig. 12–14 contd.

and cellular thickness. The largest *(L)* cross-sectional profiles (derived from juxtamedullary nephrons, most probably belonging to the longest long loops) are generally found in the vicinity of the
vascular bundles. Intermediate *(IM)* and small *(asterisk)* cross-sectional profiles of thin limbs
(derived from midcortical and superficial nephrons) are generally situated more distant from the
bundles. The venous vasa recta ascending outside the bundles originate in the inner stripe and often
lie together with thin descending limbs.

In the upper part of the inner zone (14 a) the vascular bundles are markedly reduced. The collecting ducts, however, still gather in groups of four. The thin descending and ascending limbs are
arranged randomly.

In the papilla *(14 b)* the collecting ducts have fused and have tremendously increased in cellular
height and outer diameter. The epithelium of the thin loop limbs has also slightly thickened

juxtamedullary nephrons are situated near the bundles and those of superficial nephrons together with the collecting ducts are distant from the bundles. The collecting ducts remain arranged in groups of four, thereby continuing the medullary ray pattern of the cortex.

The thin descending limbs exhibit great differences in luminal diameter and cellular thickness. The largest limbs (outer diameter up to 40 μm) are mostly situated near a vascular bundle; they belong to juxtamedullary nephrons and are believed to give rise to the longest loops, which descend into the tip of the papilla. Thin limbs of intermediate size (diameter range from 30 μm-20 μm), in their ultrastructural appearance resembling a reduced type of the large thin limbs (see, 4.2), may readily be supposed to belong to long loops of intermediate actual length; they are mostly derived from midcortical nephrons. The small thin limbs (outer diameter ～20 μm) which differ in their ultrastructural organization from the large and intermediate type (see 4.2), belong to short loops; they are mostly derived from superficial nephrons and already turn back within the inner stripe. The straight distal tubules (thick ascending limbs of Henle's loop) start either near or directly at the border between the inner and outer zones: at this site they are all equipped (regardless of whether they belong to long or short loops) with a uniformly thick epithelium that decreases toward the cortex. It decreases more quickly in juxtamedullary nephrons than in midcortical and superficial nephrons.

Inner zone (Figs. 11, 14 and 24). The inner medullary zone is characterized by a gradual vanishing of the structures that form the inner zone. The vascular bundles have already decreased when entering the inner zone (eight-ten arterial and venous vasa recta); toward the papilla they are diminished further or are absent. The long loops of Henle pass into the inner zone at variable distances and only a few reach the tip of the papilla. The collecting ducts join successively, at first within a group of four: all together they increase drastically in diameter. At the end, only a few (the number varies from two-ten giant papillary ducts) are left to open into the renal pelvis. Compared to collecting ducts, at their entry into the inner zone the diameter has increased by a factor of 6 and the cellular thickness by a factor of 10.

The architectural pattern of the upper parts of the inner medullary zone shows the collecting ducts in groups of four, still reflecting the pattern of the medullary rays in the cortex. Vascular bundles are found among them, but they can be distinguished less clearly than in the inner stripe from other structures. Thin limbs of Henle's loop are associated with both collecting ducts and vasa recta; a regular distribution is not detectable. Toward the papilla, the cross-sectional pattern is increasingly dominated by the collecting ducts; a regular histotopographic pattern among the loops and vessels is not obvious. The interstitial space and, correspondingly, the number of interstitial cells increase considerably toward the papillary tip. The volume fractions of the medullary structures and medullary interstitium in the rat and rabbit have recently been determined (Knepper et al., 1977). A comparison of the basis of the inner medulla with the tip of the papilla has shown a two-fold increase of the interstitial space in the rabbit.

3.6 Comparative Aspects of Structural Organization

Species differences in the structural organization of the renal cortex concern mainly venous drainage and the mode of formation of the collecting ducts. The different patterns of the cortical veins in mammalian species have been recently summarized by Fourman and Moffat (1971). The arrangement of the cortical veins in the rabbit follows a quite simple pattern. It consists of only radially arranged interlobular veins starting very superficially in the cortex corticis; they traverse the whole cortex accompanying the interlobular arteries and finally empty into the arcuate veins. No additional veins exist. Thus, the arrangement of the cortical veins also supports the statement that the rabbit kidney is a rather primitive kidney.

It is generally accepted (Peter, 1909; v. Möllendorff, 1930; Sperber, 1944; Oliver 1968) that there are two main types of collecting duct formation in the cortex: the collecting duct with direct junctions of individual nephrons and the arcade; both types may overlap to various degrees. In the rabbit, the juxtamedullary and deep midcortical nephrons join together into arcades before emptying into a cortical cellecting duct; the more superficial nephrons drain directly. As regards the architecture, it should be emphasized that the arcades in the rabbit have a very definite and regular histotopographic position: they ascend through the cortical labyrinth in the immediate vicinity of the interlobular vessels. Every interlobular artery has at least one accompanying arcade that runs exactly parallel to the interlobular artery and from which it is separated only by the periarterial connective tissue sheet. Surely other turbules, in most cases proximal tubules, are to be found regularly in the neighborhood of the interlobular artery, but these tubules never accompany the artery over its full length. Therefore this intimate histotopographic relationship between the interlobular arteries and the arcades seems to have not only an architectural but also a functional relevance (see 3.7.2). Nothing is known about the histotopographic relationships of the arcades in other species.

It remains a subject of controversy as to whether the direct junctions of nephrons to the collecting ducts or the junctions via arcades should be regarded the more primitive form (v. Möllendorff, 1930; Sperber, 1944; Oliver, 1968). From our point of view, i.e., that the rabbit has in many respects a primitive kidney, the arcades are considered more primitive. However, we have no concrete data to support this view.

Species differences in the architecture of the renal medulla as compared to the cortex, are of much greater interest and probably of great functional relevance. Compared to the medullary architecture of other mammalian species investigated in detail (rat: Kriz et al., 1972 a; mouse: Kriz and Koepsell, 1974; Psammomys: Kaissling et al., 1975), the rabbit renal medulla is the most simply organized.

In the rabbit kidney all the vascular bundles are of the same small size and are regularly distributed troughout a cross section of the inner stripe. They do not fuse, reflecting in the whole medulla the state of the bundles as originally developed in the outer stripe. In the other above-mentioned species, the primary bundles fuse to a different degree to form joined secondary bundles of different size (the largest ones encountered in Psammomys). The rabbit vascular bundles (as apparently in most mammalian species) consist exclusively of arterial and venous vasa recta (numerical relation, $\sim 1 : 1$); the loops of Henle are arranged roughly concentrically around the bundles. The longest loops are near the bundles; the shortest loops, most distant from them. In the other species (rat, mouse, especially in Psammomys), the descending thin limbs of short loops of Henle are integrated into the bundles white the arterial vasa recta are relatively reduced in number. This is again most apparent in Psammomys, the numerical relations between the bundle structures being about one arterial vas rectum to four venous vasa recta to four descending loop limbs (Kaissling et al., 1975). Accordingly the rabbit renal medulla may be considered to consist of uniformly developed architectural units with a vascular axis (the bundle) and the tubules arranged around it, whereas in the other species (rat, mouse, Psammomys) the architectural (and probably functional) units are of quite different size with the tubules partly integrated into the vascular axis (regarding the supposed corresponding functional differences, see 3.7.4).

Apparently related to the simple bundle organization in the rabbit renal medulla, the total venous system of the medulla is simpler than in other species. The venous vasa recta from the inner zone come together already in the inner zone to form the small innerzone bundles (together with arterial vasa recta). In the other above-mentioned species the innerzone venous vasa recta mostly ascend independently of the bundles; they do not enter the bundles before the border between the inner and outer zones. However, in all investigated species all the inner zone venous vasa recta traverse the inner stripe within the bundles.

As to the venous drainage of the inner stripe itself in all the investigated species, a large amount (most probably the majority) of the inner stripe venous vasa "recta" does not join the bundles but rather ascends directly to the outer stripe. In the rabbit a regular pattern of these venous vessels has become apparent. Recalling that within the interbundle regions the collecting ducts descend in groups of four, it becomes obvious from microfil-injected specimens that the venous vasa recta avoid the collecting duct areas ascending between these areas as well as between them and the vascular bundles. A cross section through the inner stripe exhibits a network pattern of the venous vasa recta; through its holes descend the collecting duct groups. Thus, the interbundle venous vasa recta are associated with the loops of Henle, which are situated in the areas between the collecting duct groups and the vascular bundles; at this site the descending thin limbs predominantly contact the ascending venous vasa recta. Similar histotopographic relationships are found in other mammalian species for which – as in the rabbit – no integration of thin limbs into the vascular bundles exists, e.g., in the cat (Kriz, 1970).

The venous pattern of the outer stripe (all venous vessels traverse the outer stripe as wide capillary channels contacting the tubules) is similar organized in all mammalian species (investigated so far). However, the further route of the venous vasa recta differs. In the rabbit they all empty at the corticomedullary border into arcuate or interlobular veins (basal parts). In contrast, in the rat or in Psammomys many venous vasa recta ascend a considerable distance within the medullary rays and finally empty into middle or even upper parts of interlobular veins.

In summary, the simple medullary architecture of the rabbit kidney is considered to be a primitive one. The more complex patterns in other species (rat, mouse, Psammomys) represent a more advanced stage with higher functional efficiency.

3.7 Functional Aspects of Structural Organization

This chapter will summarize the possibilities of functional interaction between different structures of the kidney as are obvious from their histotopographic relationships.

3.7.1 Interstitial Space

Both the cortex and medulla of the kidney have their own interstitial compartment in which tubules and vessels are embedded. With the exception of the periarterial loose connective tissue sheet, the cortical interstitium is sparse, while the medullary interstitium becomes more and more abundant toward the tip of the papilla (Dieterich and Kriz, 1972; Bulger and Nagle, 1973). From the morphologic organization it can be deduced that direct communications between the cortical interstitial fluid and the medullary interstitial fluid are negligible. At the border between cortex and outer stripe (a border which cannot be defined exactly), there are interstitial communications to which, at best, merely local importance may be ascribed. The interstitial compartment within the medulla is subdivided into an inner and outer medullary compartment. Only local importance is ascribed to the direct communications at the common border.

The cortical interstitial compartment has lymphatic drainage, while the medulla has none. In the cortex the surplus of interstitial fluid (the largest portion is believed to be reabsorbed into the peritubular capillaries) is expected to leak via some type of convective flow toward the interlobular and arcuate arteries. The periarterial loose connective tissue sheets of the latter provide the base for a hilus-directed interstitial fluid flow. As the lymphatics of the kidney are embedded in the periarterial tissue, the interstitial fluid may be successively reabsorbed into the lymphatics and drained off. The medulla has no lymphatic drainage; therefore, the total quantity of generated medullary interstitial fluid must be taken up and drained off by capillaries and venous vessels.

Within these interstitial compartments (cortex, outer medulla, inner medulla), neighboring structures, tubules, and vessels are functionally connected by the common interstitial fluid. Functional interactions between tubules and adjacent capillaries may even be more direct at those places where, as frequently happens in the cortex, the capillary and tubular basement membranes are closely apposed (Bulger and Nagle, 1973). It is difficult to judge the functional significance of the direct appositions between tubules and capillaries. Neither the amount of outer tubular surface directly related to capillaries nor the amount of outer capillary surface directly related to tubules is known. As regards small molecules, the peritubular capillaries are extraordinary permeable (Barger and Held, 1973). Equilibrium between plasma and interstitial fluid composition may therefore be expected; thus, for small molecules the functional relevance of these direct appositions does not differ from that via the interstitial space. For large molecules (e.g., albumin), equilibrium cannot be expected; the albumin concentration in the interstitium is estimated to be in the magnitude of one-third of the plasma concentration (Pinter, 1977). Accordingly, for these molecules, related fluid flows, and ion distribution patterns, the direct appositions could be of functional significance.

3.7.2 Possibilities of Influence on the Arterial Input

The intrarenal arterial blood flow may be influenced at the following sites: arcuate and interlobular arteries, vas afferens, vascular pole, vas efferens, and (only valid for the medulla) arterial vasa recta.

Influence possibilities on the arterial input are given by (a) the outward directed flow of the interstitial and lymph fluid, (b) the capillary and venous blood flow, and (c) the tubular urine at direct contact sites (or narrow histotopographic relationships) between tubules and arterial vessels.

Interstitial fluid and lymph flow (Fig. 15 a). The interstitial fluid of the cortex is expected to be generated from capillary blood, from tubular reabsorbate, and in addition from a filtrate of intrarenal veins (Bell at al., 1970; Vogel et al., 1974). It is obvious that an interstitial fluid within the cortical parenchyma acts directly on the vas efferens, the vascular pole, and the vas afferens. It can be assumed that a surplus of interstitial fluid gains access to the periarterial loose connective tissue; there it might directly act upon the smooth muscle cells of the vas afferens, interlobular and arcuate arteries. The lymphatics embedded into this loose connective tissue must be regarded as a part of this hilus-directed drainage system. Substances (e.g., kallikrein, renin) that have been taken up by the lymphatics (via the periarterial tissue) could possibly leak out of the lymphatics at sites nearer to the hilus. They may then act on

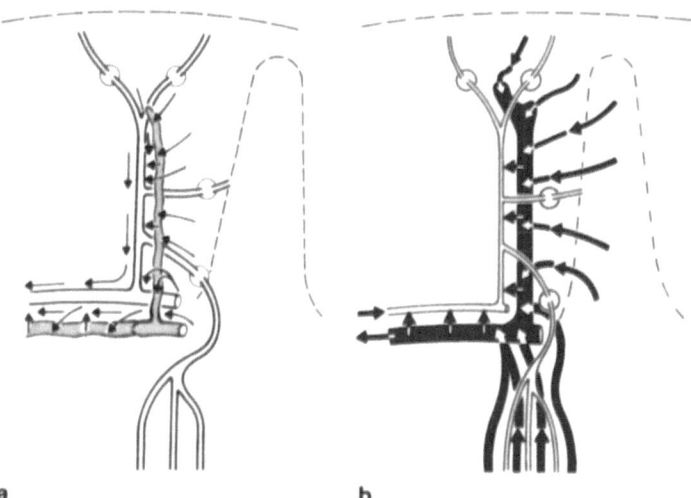

a   b

Fig. 15 a and b. Possibilities of influence to the arterial input. *a.* Interstitial fluid and lymph flow. *Arrows* indicate that a surplus of interstitial fluid from the cortical parenchyma can be expected to gain access to the periarterial loose connective tissue and the lymphatics (gray). Lymphatics and periarterial connective tissue establish a common drainage system by which interstitial fluid can leave the kidney. Since blood flow in the arteries and interstitial fluid/lymph flow are countercurrent, substances carried with interstitial/lymph fluid may act on the wall of all intrarenal arteries. Interstitial fluid from the medulla is excluded from this route.*b.* Venous blood flow. Veins and venous vasa recta are *black.*
Arrows indicate that substances leaving the kidney via the vein may (on their countercurrent route toward the hilus) leave the veins (capillary wall!) and act upon the arterial wall. In contrast to the interstitial/lymph fluid route *(a)* also substances originating from the medulla may act via the venous blood flow on renal arteries

more proximal portions of the arterial tree (arcuate arteries or even interlobar arteries). Renin is expected to be secreted into the interstitial space at the end of the vas afferens (Johnston et al., 1975), i.e., into the beginning of this drainage system. In addition, this drainage route might also be of importance for the action of kallikrein, which is assumed to be synthesized in some part of the distal tubule within the cortex (Orstavik et al., 1976).

Capillary and venous blood flow (Fig. 15 b). The two cortical capillary plexuses are functionally connected in that blood from the medullary ray plexus must pass the labyrinth capillaries to gain access to the interlobular veins. On this route substances originating from the medullary rays might act on those parts of the arterial tree which are situated within the labyrinth (vas efferens, vas afferens).

The interlobular veins run parallel to the corresponding arteries as do the arcuate and interlobar veins. According to their wall structure, interlobular and arcuate veins (consisting only of fenestrated epithelium) must be regarded as capillaries (Dieterich, 1978; Dieterich investigated the intrarenal veins in the rat; we found the veins in the rabbit equivalent to those of the rat). Substances in the venous blood might leave these veins, gain access to the periarterial tissue, and thus to the arterial smooth muscle cells.

The venous blood of the medulla ascends within the venous vasa recta. These vessels have close contact with the arterial vasa recta and the vasa efferentia of juxtamedullary nephrons. By this arrangement the blood flow to the medulla may be influenced independently from the cortical blood flow.

Finally, the medullary venous blood drains into the arcuate veins, which are also capillary-type vessels (Dieterich, 1978). Thus, substances originating from the medulla (e.g., prostaglandins) might diffuse from the arcuate veins as well as from basal parts of interlobular veins and come into contact with the smooth muscle sheet of the arcuate arteries. In other species (e.g., rat, Psammomys) the medullary venous vessels partly join the interlobular veins far up in the cortex (Rollhäuser et al., 1964; Bankir et al., 1978). Therefore in these species substances originating from the medulla might reach the interlobular arteries. In the rabbit, this possibility is restricted to the basal parts of the interlobular arteries.

Urine flow. The contact field of the ascending limb to the vascular pole of the corresponding glomerulus, the macula densa, is obviously the most important feedback site for the intrarenal regulation of blood flow. In this context it needs no further discussion.

A second contact field has been frequently (if regularly is not known; see 3.4.3) found between the pre- or postmacula segment of the straight distal tubule and the vas efferens. It seems possible that here, independently from the juxtaglomerular apparatus (JGA), the vas efferens is influenced by the composition of the distal tubular urine. However, even morphologic evidence needs further investigation.

A third possibility for a tubular arterial input feedback could be established by the arcades. In the rabbit, the arcades ascend within the cortical labyrinth to the interlobular arteries in a conspicuously parallel manner. They do not really touch the arteries, but contact the periarterial tissue. Thus, substances released from the arcades (e.g., kallikrein? prostaglandins?) need only traverse the periarterial layer of loose connective tissue to reach the interlobular arteries and the basal parts of the vasa afferentia. No data prove if this conspicuous histotopographic arrangement is of any functional relevance.

To summarize, as regards the morphologic arrangement, much indicates that not only the arterioles are responsible for the intrarenal regulation of blood flow (vas afferens, vas efferens) but also the intrarenal arteries (interlobular and arcuate arteries). Källskog and co-workers (1976) recently presented convincing functional evidence supporting the possibility that interlobular arteries make an important contribution to the regulation of intrarenal blood flow.

3.7.3 Possibilities of Influence on the Venous Output

The intrarenal venous vessels (venous vasa recta, interlobular and arcuate veins) are, due to the structural organization of their wall, capillary like vessels (Dieterich, 1978). Accordingly they may be expected to have permeability properties similar to those of renal capillaries. Moreover, their delicate wall, which lacks sufficient connective tissue support, appears unable to stabilize the form of the vessel, and their wall cannot actively resist intravascular pressure because of its lack of smooth muscle sheet. Thus, the lumen of these venous vessels is completely dependent on the hydrostatic pressure difference between inside and outside.

The intrarenal venous pressure has been reported to be relatively high and to drop abruptly somewhere along the arcuate or interlobar veins (Aukland, 1976). As there is no structural basis for assuming an "active venous resistance," the capacity of the venous vascular bed in total may be interpreted as relatively small and the total resistance as relatively high. The abrupt pressure drop would indicate that the total resistance in the large veins leaving the renal parenchyma (interlobar veins) suddenly decreases. This would be in accordance with the morphology in as far as the resistance of the intrarenal veins (interlobular veins) the walls of which are fully lacking in structural rigidity, must depend decisively on the intrarenal (interstitial) pressure).

When the veins leave the renal parenchyma and run as interlobar veins within the loose connective tissue of the side walls of the renal pelvis, the pressure outside the veins might well be smaller. Thus, in the kidney it may be assumed that changes in the intrarenal (interstitial) pressure will act on the intrarenal venous resistance. This interpretation agrees with the fact that elevation of intrarenal pressure by increased ureteral pressure or by diuresis increases pressure in the intrarenal veins (Aukland, 1976).

As regards the renal medulla, the possibilities of influencing the venous output have additional aspects. The total number of medullary venous drainage vessels traverse the outer stripe as wide capillary channels. Changes in the intrarenal (interstitial) pressure in the outer stripe of the renal medulla, which may to some extent change independently of other parts of the kidney, may be expected to strongly influence the blood flow in these vessels, i.e., the total venous output of the medulla. Thus, differences in the distribution of blood flow between cortex and medulla may also be based on different resistances to the venous output.

3.7.4 Vessel Bridges Between Different Tubular Portions – Recycling Routes

A discussion of tubulo-/tubular- interaction possibilities should show first which interactions are not possible. In the cortex as well as in the medulla no direct contacts between two tubules occur. Indeed, in the cortex two tubules can come very close to each other; yet, a fusing of two corresponding basal laminae is seldom to be observed.

Therefore, in all regions of the kidney interactions even between neighboring and adjacent tubules are only possible via the common interstitial space. For example, it has often been discussed whether the descending and the ascending limb of Henle's loop directly influence each other. The single nephron injections (see 3.4.3) demonstrate that both corresponding limbs of a loop generally run next to and parallel to each other over long distances. Nevertheless, interaction between descending and ascending limbs is only possible in such a way that the total number of descending limbs of a given area acts via the interstitial space on the total number of ascending limbs (and vice versa). An individual descending limb cannot precisely and exclusively act on the corresponding ascending limb.

All interaction possibilities between distant nephron portions are given by vessel bridges. These bridges can be described within the cortex and most importantly within the medulla, but not — at least in the rabbit kidney — between the cortex and the medulla.

The vessel bridge within the *cortex* has already been mentioned. Since the venous blood from the medullary ray capillaries passes through the labyrinth capillaries to gain access to the interlobular veins, a back connection from the straight tubule in the medullary rays (straight proximal and straight distal tubules of superficial and midcortical nephrons as well as cortical collecting ducts) to the convoluted tubules in the cortical labyrinth (proximal and distal convoluted tubules) is established. This arrangement provides the basis for a cortical recycling system, by means of which substances reabsorbed from the straight tubules and cortical collecting ducts are reoffered to the convoluted tubules.

In the *medulla* the vascular pattern establishes remarkable functional connections as well as separations between different tubular portions. To understand them correctly, one should imagine that each of the three medullary regions (outer stripe, inner stripe, inner zone) has its own supplying vessels (i.e., arterial vasa recta or in the outer stripe direct branches from juxtamedullary vasa efferentia) and its own drainage vessels (i.e., venous vasa recta, except in the outer stripe). The direct capillary connections between the capillary plexuses of the three regions are considered to be at the most of merely local importance. Thus, the obvious functional connections and separations between the three medullary regions are established exclusively by the vasa recta.

The arterial vasa recta intended for the inner zone traverse the inner stripe within the bundles. Therefore blood reaching the inner zone has not previously been in contact with tubules of the inner stripe. All the venous vasa recta originating from the inner zone traverse the inner stripe within the bundles and the blood that has been in contact with the tubules of the inner zone never contacts the tubules of the inner stripe. The blood flow of the inner stripe and that of the inner zone are apparently separated from each other.

Within the vascular bundles of the inner stripe, the venous vasa recta coming from the inner zone establish a countercurrent arrangement (Fig. 16 a) with the arterial vasa recta, which are partly determined for the inner zone and partly for the inner stripe. Therefore substances (e.g., urea) originating from the inner zone could be transferred from the ascending venous vasa recta to the descending arterial vasa recta. On this path they would partly return to the inner zone and partly be shifted to the inner stripe capillary plexus where they may be offered to the inner stripe tubules (thin descending limbs, thick ascending limbs). In species containing complex vascular bundles (i.e., an integration of the thin limbs of short loops into the bundles; e.g., in

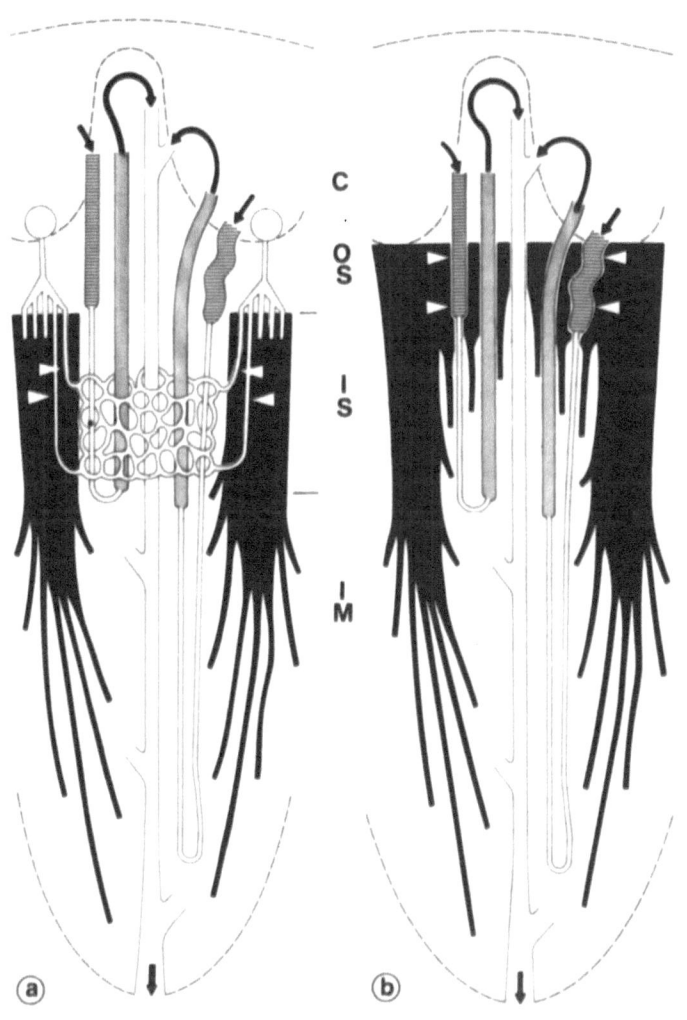

Fig. 16 a and b. Medullary recycling routes. Arterial vessels, thin loop limbs, and collecting ducts are *white*, venous vessels, *black*, straight proximal tubules are *hatched*, and straight distal tubules are *gray*. *a.* Recycling possibility at the level of the inner stripe. Within the vascular bundles of the inner stripe, venous vasa recta ascending from the inner medulla run countercurrent to arterial vasa recta, which supply the inner stripe. Accordingly substances from the inner medulla can be transferred *(white arrow heads)* to the inner stripe capillary plexus. *b.* Recycling possibility at the level of the outer stripe. Ascending venous vasa recta from the inner medulla and the inner stripe traverse the outer stripe in close contact to the tubules, running countercurrent to straight proximal tubules. Substances from the venous vessels can be transferred by secretion into proximal tubules (S3 segments!) and thereby return to the medulla or reach more distal nephron portions. *C* = cortex; *OS* = outer stripe; *IS* = inner stripe; *IM* = inner medulla

the rat, mouse, Psammomys), the possibility for some substances (e.g., urea) originating from the inner zone to be transferred to the thin descending limbs of short loops is much more direct, since the venous vasa recta ascending from the inner zone and the descending limbs of short loops are arranged in a countercurrent exchange system (Kriz et al., 1976).

This interpretation of inner-stripe architecture — irrespective of whether in the rabbit or in other species — is based on the assumption that within the vascular bundle area and the interbundle region differences in the interstitial fluid composition could be maintained. This certainly may apply to species with giant bundles (Psammomys, Meriones); in the rabbit such a separation is less obvious, but may also be supposed. The inner stripe can be regarded as composed of two different interstitial compartments: the bundles and the interbundle regions. In contrast, tubules and vessels within the inner zone are embedded into the common interstitial space.

The tubules of the outer stripe are mainly supplied by the ascending venous vessels. Due to this arrangement, the total medullary venous blood originating from the inner zone and the inner stripe comes into contact with the tubules of the outer stripe (Fig. 16 b), predominantly with the straight proximal tubules. A loose subdivision could be based on the fact that vasa recta spreading out from the bundles predominantly contact the proximal tubules of juxtamedullary nephrons, whereas venous vasa recta ascending directly from the inner stripe predominantly contact straight tubules of midcortical and superficial nephrons.

These striking vascular relationships in the outer stripe may be important for several mechanisms (Fig. 16 b). It can be regarded as the final possibility for preventing the renal medulla from losing osmotic energy via the venous blood. Provided that the osmotic concentration in the venous blood at this level is still above plasma concentration, even a small gradient could be utilized by proximal tubules (elevating the iso-osmolar reabsorption to a level somewhat above plasma concentration) and the collecting ducts. Moreover, these vascular tubular relationships could play a major role in trapping a substance within the medulla. In this context it is significant that we are concerned with end portions of the straight proximal tubules (S3 segments) which are known to be extraordinarily qualified for secreting several classes of substances into the tubular lumen (Grantham and Irish, 1978). The possibility of recycling into the medulla is offered to all these substances. Should they leave the loop of Henle within the medulla, recycling along this route might be perpetuated. Substances originating in the medulla (e.g., prostaglandins) may also reach more distal nephron portions by this route.

Since these relationships between the medullary venous vessels and descending tubules have been found in all mammalian species examined, it must be of great functional relevance. However, the extent of these venous vasa recta — descending tubule — relationships do differ according to species. In the rabbit, all venous vasa recta empty at the corticomedullary border into arcuate veins or into the basal most parts of interlobular veins. Thus, the trapping possibilities are restricted to the outer stripe. In other species (e.g., the rat, Psammomys) the medullary venous vessels ascend for a considerable distance within the medullary rays. They thereby extend the contact with the descending proximal tubules far up into the cortex. In this regard, the pattern in the rabbit also appears to be more primitive and less effective.

3.7.5 Recycling Possibilities from the Renal Pelvis

It has frequently been discussed whether or not the histotopographic relationships of the renal pelvis offer possibilities for a back diffusion of solutes from the pelvic cavity into the renal parenchyma. A correlation has been suggested between the degree of the development of pelvic protrusion into the medullary parenchyma and the relevance of urea in the concentrating mechanism (Pfeiffer, 1968; Schmidt-Nielsen, 1977).

The renal pelvis in the rabbit envelops the total inner medulla and with its secondary pouches broadly contacts the inner stripe of the outer medulla. Accordingly, a back diffusion of solutes (e.g., urea) from the pelvic cavity into the renal parenchyma may be directed into the inner medulla and to a considerable degree also into the inner stripe. The rabbit renal pelvis, however, has no tertiary protrusion toward the vascular bundles into the inner stripe parenchyma as they are extensively developed in Psammomys (Kaissling et al., 1975).

Moreover, a solute diffusion leakage out of the pelvic cavity might also be directed into the pelvic septa and then into the interlobar veins and interlobar lymphatics. Whether or not this possibility is of any physiologic relevance ist not known; the composition of the hilar lymph might, however, be considerably determined by such an input. In saline diuresis the hilar lymph contains higher concentrations of Na^+ and Cl^- than the plasma (O' Morchoe et al., 1978), perhaps easily due to an addition from the pelvic urine.

3.7.6 Nerves

An investigation of the distributional pattern of nerves in the rabbit kidney has not yet been performed. From the literature (Gosling, 1969; Gosling and Dixon, 1969; Norvell et al., 1969; Müller and Barajas, 1972; Barajas et al., 1974; Barajas and Wang, 1975; Dieterich, 1974; Gorgas, 1978) the innervation pattern of the kidney in various mammalian species is known to be very similar.

The nervous apparatus in a mammalian kidney consists of nerves and terminal axons (containing varicosities) which accompany the renal arteries and arterioles (vas afferens, vas efferens, arterial vasa recta as long as they are enveloped by smooth muscle cells). Thus, no nerves or terminal axons penetrate into the renal parenchyma. Tubules may have direct histotopographic relationships to terminal axons only if they are situated in the vicinity of arteries or arterioles (Gorgas, 1978).

Disregarding the question as to the type of nerves (exclusively sympathetic or in addition parasympathetic and afferent nerves), we briefly discuss the problem of whether an "innervation" of tubules at distance is regularly possible. The possibility of a direct neurogenic influence on tubular sodium reabsorption is considered controversial (di Bona, 1977).

As may be assumed, nerve fibers pass along the vascular pole of each glomerulus from the vas afferens to the vas efferens (Gorgas, 1978; Doležel et al., 1976). Therefore the overall density of nerves in the renal cortex is considerable. Substances (e.g., catecholamines) released from these nerves, especially from those at the vascular pole and the vas efferens, should gain access to the peritubular capillaries of the cortical labyrinth via the interstitium. As the blood flow in these capillaries is expected to be predominantly directed toward the interlobular veins, the convoluted tubules of the cortical labyrinth might regularly be reached by substances of the capillary blood flow. However, quantitative estimation (percentage of axons contributing to this possibility) is not possible. Action on the straight tubules within the medullary rays of the cortex is less probable, since the capillary blood flow is not expected to go in the appropriate direction.

The close histotopographic relationships of the arcades to the interlobular arteries might also be of functional importance as regards the nerves. As the latter accompany the arteries, released catecholamines might easily reach the arterial smooth muscles and the arcades (connecting tubules). In this context it has to be mentioned that Morel and co-workers (1976) have shown that the connecting tubules in the rabbit kidney are sensitive to isoproterenol.

In the medulla the nerves descend some distance within the vascular bundles (in the rat, they do not reach deeper levels than the upper part of the inner stripe; Dieterich, 1974). Substances released from these nerves can be expected to gain access into the bundle venous vasa recta. Via these vessels they might reach the straight tubules (predominantly those of juxtamedullary nephrons) of the outer stripe. A quantitative estimation and thus a judgement of the relevance of this route is impossible.

4 Ultrastructural Organization of Nephrons and the Collecting Duct System

The ultrastructural organization of the various tubular portions in the mammalian kidney is known to varying degrees. The proximal tubule has been very intensively investigated in various species (Rhodin, 1958; Sakaguchi and Suzuki, 1958; Thoenes, 1961 a, b; Thoenes and Langer, 1969; Maunsbach, 1966, 1973; Tisher et al. 1966; Tisher, 1976). In the rabbit, however, a systematic investigation with advanced techniques is lacking. As the ultrastructural organization is similar to other species, only a relatively short description will be given.

Since the thin limbs of the loop of Henle and the distal tubule have been insufficiently described and moreover considerable species differences complicate research, we have investigated both of them in detail.

On an ultrastructural level the transition from the distal tubules to the collecting duct system is not clearly understood. In the rabbit a distal tubule was found always linked to a cortical collecting duct by a cytologically distinct tubular portion. For several reasons (see 4.4.3) we have called this portion the connecting tubule. Due to similarities in their ultrastructural organization the connecting tubule is considered part of the collecting duct system. The ultrastructure of the collecting ducts of the rabbit kidney also required systematic investigation.

4.1 Proximal Tubule Ultrastructure

The tubular system of the rabbit nephron starts with a tubelike elongation of the outer leaflet of Bowman's capsule of the renal corpuscle and forms a so-called neck segment, as described in detail by Schønheyder and Maunsbach (1975). This neck segment can be of considerable length but can also be virtually absent. No correlation was found between the length of the segment and the nephron type (superficial, midcortical, or juxtamedullary).

The proximal tubule (Figs. 17, 18 and 19) of the rabbit gradually alters its ultrastructural appearance over its full length. Since a comparison of the cells in the beginning and end portion reveals rather different cell types, it is reasonable to arbitrarily distinguish three parts in the rabbit proximal tubule as in other mammals (Maunsbach, 1973; Jacobsen, 1975). Segment 1 (S1; Fig. 17) emerges abruptly from the neck segment and comprises the initial and middle part of the proximal convolution (lying exclusively in the cortical labyrinth). Segment 2 (S2; Fig. 18 a) roughly extends from the late part of the proximal convolution to the first half of the straight proximal tubule. Segment 3 (S3; Fig. 18 b) comprises the rest of the straight part (lying within the medullary rays and the outer stripe of the medulla). S2 can be considered a transitional segment between the very elaborate cells in S1 and the much more simply organized cells in S3; neither the beginning nor the end of segment S2 can be clearly delineated. The alterations in the ultrastructural organization of the proximal tubule concern nearly all characteristics of a proximal tubule cell. Thus, the alterations for each characteristic will be described separately.

The height of the cell body (microvilli not included) gradually decreases from ~10 μm in the first segment to ~5 μm in S3 with the steepest decrease along S2. The long and slender microvilli (which typically contain seven-ten longitudinally arranged microfilaments apparently anchored in the cytoplasm) are most densely packed and longest in S1 with a length of ~3 μm (nearly one-third of the height of the cell body). Toward S3 they decrease in length to ~2 μm in S3 itself. In relation to the height of the cell body in S3, they appear to be longer. The microvilli often, most obviously in S3, originate in bundles from a pedestallike protrusion of the apical cell membrane.

Laterally and basally the cells of the proximal tubule interdigitate with complexly shaped processes. Basally these processes are further split into small basal villi (which typically contain a condensation of filamentous material at their base). In S1 the interdigitation of cells is most elaborately developed; the cell processes are regularly arranged in the apical basal direction extending over almost the full height of the cell body. Already in S2, the interdigitated part is restricted to the basal two-thirds of the cells. In this segment the basal ramifications (basal villi) are most extensively developed and create a regular basal rim which is fully devoid of cell organelles (Fig. 18 a). In S3 the interdigitating processes are irregular and restricted to the basal most regions of the cell.

Proximal tubular cells are richly stuffed with mitochondria. Their shape and arrangement depends mainly on the interdigitating cell processes. Accordingly, the very numerous and large mitochondria in S1 are rod-shape and lie within the lateral cell processes. Corresponding to the gradually altering forms of interdigitating cell processes (toward and within S2), the mitochondria lose their parallel arrangement and become more randomly distributed. They no longer have a rod-like shape but often appear tortuous and smaller. In S3 the mitochondria lie randomly in the cytoplasm without association to lateral cell membranes.

Golgi apparatus and endoplasmatic reticulum are well developed in proximal tubular cells. The smooth form of the endoplasmatic reticulum predominates, but profiles of granulated endoplasmatic reticulum and polysomes are also amply present.

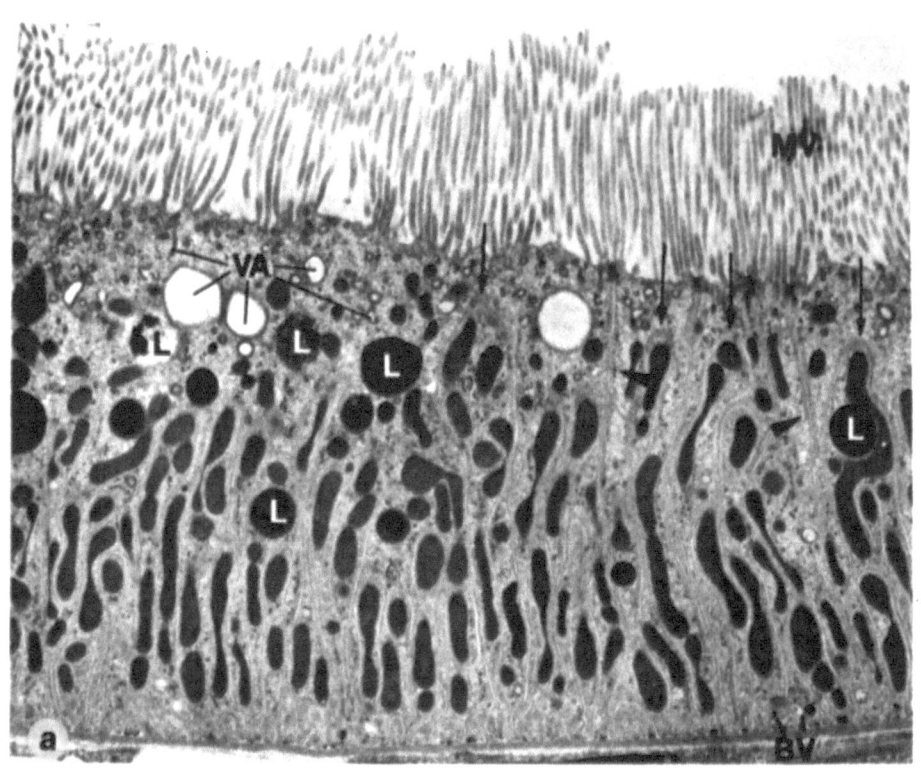

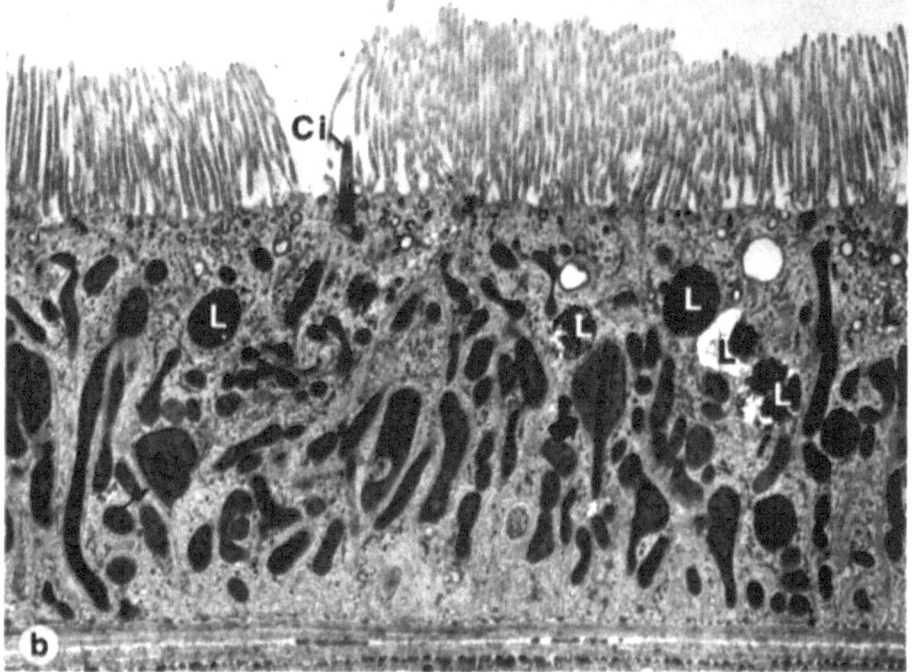

Fig. 17

Transitions from one form to the other can frequently be found. In S1 (apart from the paramembraneous tubular system; see below) the endoplasmatic reticulum is rather concealed, since the cells are densely stuffed with other cell organelles. Within S2 the smooth-surface endoplasmatic reticulum becomes exceedingly prominent and exclusively fills large areas of the cytoplasm. In S3 granular profiles become increasingly apparent and transitions between the smooth and the granulated forms can frequently be observed. A special formation of the endoplasmatic reticulum is the paramembraneous tubular (or cisternal) system (PTS or PCS). It consists of long flat saccules extending along the lateral cell walls. Generally the PTS is made of smooth reticulum, but transitions to the granulated form occur frequently, According to the reduction of the lateral cell membrane interdigitations, the PTS greatly decreases along the proximal tubule.

The prominent lysosomal system of the proximal tubule typically consists of the so-called vacuolar apparatus and a high amount of irregularly distributed multiform lysosomes. The vacuolar apparatus situated in the apical cytoplasm consists of small endocytotic vesicles, large vesicles, and many tubelike vesicles. All have a conspicuous membrane with an amorphous coat on the outer and inner face. In S1 the small vesicles dominate the vacuolar apparatus, whereas in S2 and even more conspicuously in S3, the content of large vacuoles increases. The irregularly distributed multiform lysosomes increase in number from S1 to S3, whereby the amorphous internal substance in S1 generally stains dark and then becomes lighter in S2 and S3.

Peroxisomes are a typical organelle of proximal tubular cells because they are exclusively present in this nephron portion. They are predominantly found in the basal and paranuclear regions of the cell and in S2 also in the apical cytoplasm. In S1 of superficial and midcortical nephrons, peroxisomes are rarely present. In juxtamedullary nephrons large and conspicuous peroxisomes are already numerous in S1. They differ from those in subsequent nephron portions, since in addition to marginal plates they exhibit tubular profiles and so-called phi bodies (Hanker et al., 1977). In S2 and S3 (of all nephrons) peroxisomes are quite numerous; they show marginal plates and their content is generally amorphous. Peroxisomes are often closely related to profiles of smooth endoplasmatic reticulum, which may fully surround a peroxisome and come in close proximity of the marginal plates.

Differences in the nuclei of the proximal tubule are not obvious. In all segments they have conspicuous heterochromatin condensations, are often partially lobulated, and have a prominent nucleolus, occasionally two nucleoli.

Fig. 17 a and b. Proximal tubule. *a*. Cells of the first segment (S1) of superficial nephrons. Interdigitating cell processes in parallel arrangement contain long slender mitochondria and extend into the apical cell region *(arrow)*. Dark staining lysosomes *(L)* are amply present; the vacuolar apparatus *(VA)* consists of few large vacuoles and many small endocytotic vesicles and tubules; the microvilli *(MV)* of the brush border are long. *Arrow heads* = paramembraneous tubular system (PTS); *BV* = basal villi. *b*. Cells of S1 of juxtamedullary nephrons. The general organization is similar to that in *(a)*; yet, in contrast, large peroxisomes with cristalline internal structure *(thin arrow)* and phi-bodies *(thick arrow)* occur. Cilia *(Ci)* can be present on each cell. *(a)* and *(b)* x ~6800

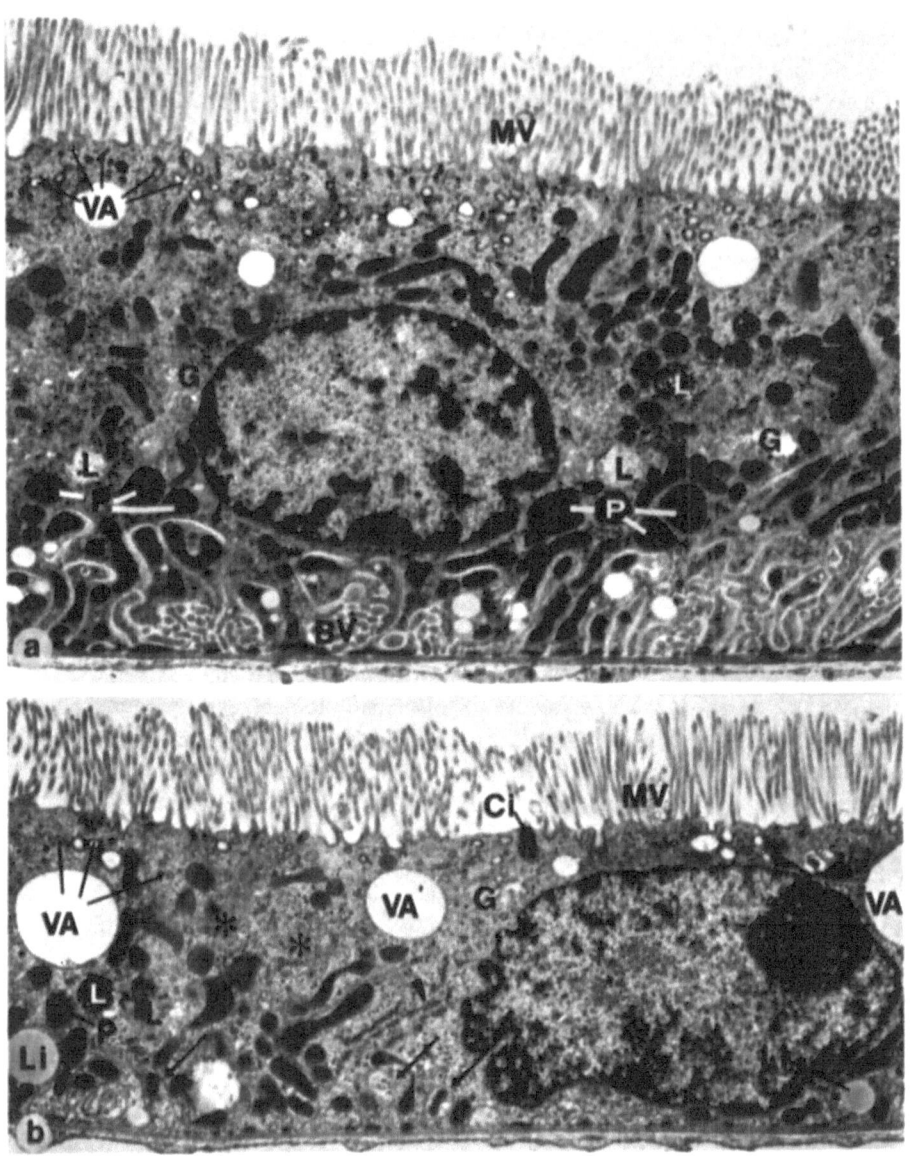

Fig. 18 a and b. Proximal tubule. *a.* Cells of the second segment (S2) in the medullary ray. The interdigitation of the cells is restricted to the basal cell half *(thin arrows)*. Mitochondria are distributed randomly; lysosomes *(L)* are less dark staining than in S1 and peroxisomes *(P)* with marginal plates are amply present. A large rim of basal villi *(BV)* is conspicuous in this segment; the brush border *(MV)* is shorter than in S1; *G* = Golgi apparatus; *VA* = vacuolar apparatus. *b.* Cells of the third segment (S3) in the outer stripe. The height of the cell body has decreased, whereas the microvilli of the brush border *(MV)* are as long as in S2. Only few small interdigitations are present *(thin arrows)*. Between the small number of mitochondria are prominent large areas occupied by smooth endoplasmatic reticulum *(asterisks)*. Cisternae of rough endoplasmic reticulum (arrow head) occur. Vacuoles of the *VA* can be very large, whereas small endocytotic vesicles decrease in number. Lysosomes *(L)*, peroxisomes *(P)*, as well as lipid droplets *(Li)* are regularly present. *G* = Golgi apparatus, *Ci* = Cilia, *VA* = vacuolar apparatus. *(a)* and *(b)* x ~6800

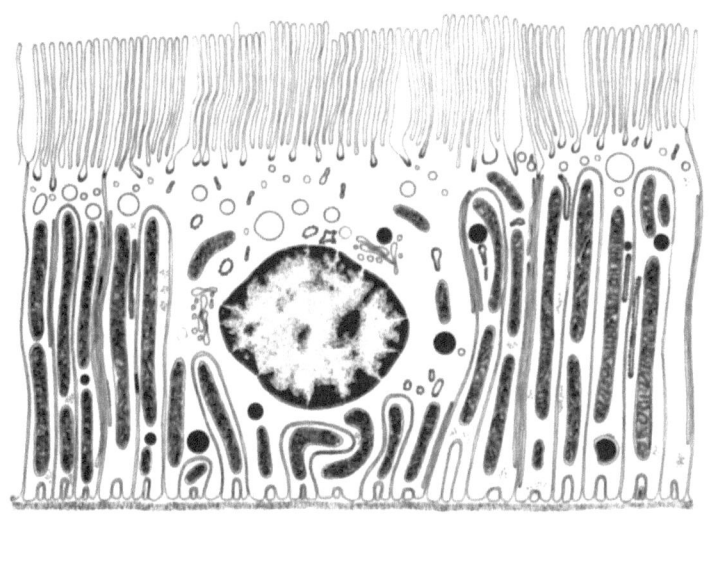

(a)

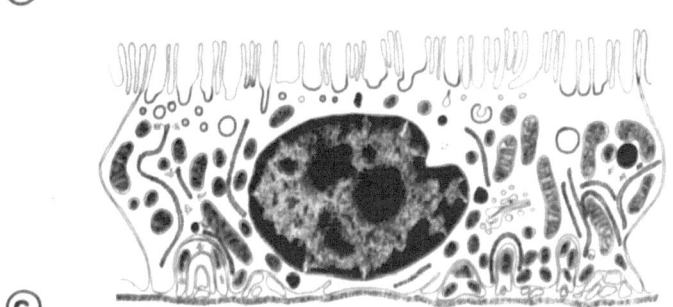

(b)

(c)

Fig. 19. Proximal tubular cells in *(a)* segment 1, *(b)* segment 2, and *(c)* segment 3. x ~4500

Apically the proximal tubular cells are joined by shallow tight junctions (one-three ridges); differences within the various segments are not apparent. This finding corresponds to freeze-fracture data (Roesinger et al., 1978), in which a uniform tight junctional belt consisting of one-three ridges was found in all portions of the proximal tubule. Immediately beneath the tight junctional belt a deep zonula adherens is present. Desmosomes, if present, are only small. Gap junctions between the lateral cell processes are regularly and most frequently found in S1.

The basal lamina varies in thickness along the proximal tubule. In S1 and most obviously in S2 it is conspicuously thick and exceeds the basal lamina of the distal tubules or of the collecting ducts by three to four times. However, in S3 the basal lamina decreases to the thickness of other nephron portions.

4.1.1 Comparative and Functional Aspects

Histologically the proximal tubule is subdivided into a convoluted and a straight part. A more detailed subdivision has been proposed by Zimmermann (1911; in the dog and cat) which principally corresponds to the subdivision into three segments based on ultrastructural criteria (Maunsbach, 1966, 1973; Ericsson and Trump, 1964, 1966; Ericsson et al., 1965; Beard and Novikoff, 1969). In the rabbit, however, the transitions between the segments are gradual, making segmentation a matter of discretion. Both subdivisions have advantages and disadvantages; convoluted and straight parts may be easily distinguished by their localization. Ultrastructurally, however, each consists of different cells. Therefore, when comparing the structural with cellular function, a more precise subdivision is necessary. Optimally the proximal tubule should be divided in as many segments as possible, but to avoid confusion the subdivision into three segments represents a good compromise.

Although the overall morphology of the proximal tubules seems quite similar in different mammalian species (Maunsbach, 1973) a closer examination may well reveal differences. In the rat (Maunsbach, 1973) and dog (Zimmermann, 1911), the transition from S2 to S3 has been reported to be rather abrupt; S1 feline cells contain a high amount of large lipid droplets (Bargmann et al., 1977; Schiebler, 1959); S3 in the rat is characterized by very long (the longest in the whole proximal tubule) microvilli. Species differences are also found in the tight junctional belt. Whereas in the rabbit the tight junctional belt does not alter significantly along the proximal tubule (consisting only of one-three parallel strands), in other species (the cat, Tupaia, dog, rat; Roesinger et al., 1978), the number of strands and the apical basal depth increase considerably within the straight part toward S3. Those differences (alterations in the quality of paracellular pathway) may well be parallelled by functional differences.

Although direct correlations between ultrastructure and function should only be drawn with caution and are still fully impossible with regard to intrinsic membrane characteristics, they may well aid basic comprehension of cellular organization. Gradual functional alterations along the proximal tubule have been shown in physiologic experiments in which the proximal tubule has been subdivided into more than two or three segments (Ullrich et al., 1977: phosphate transport). Thus, gradual functional alterations may be expected also to become apparent in those functions investigated by polar experiments, i.e., by comparison of the convoluted with the straight part (Tune and Burg, 1971: glucose transport; Bourdeau and Carone, 1974: albumine uptake; Kawamura and Kokko, 1976: urea secretion; Ashton and Koepsell, 1976: Na–K–ATPase; Knox et al. 1977; Greger et al., 1977: phosphate transport). The most striking functional alteration along the proximal tubule is the apparently gradual decrease in the salt reabsorptive capacity (Burg and Orloff, 1968; Schafer et al., 1977; Frömter and Geßner, 1974 a, b; Burg, 1976), which ultrastructurally corresponds to the decrease in structural features thought to be primarily involved in salt reabsorption, i.e., microvilli, lateral basal interdigitations, and mitochondria. Investigations by Welling and Welling (1975, 1976) have demonstrated (by comparing S1 with S3) that the reduction of luminal cell surface area quantitatively parallels the reduction of the lateral and basal cell surface area. In S1 both the luminal and lateral surface are enlarged about 36 times (by microvilli and lateral cell processes, respectively); in S3 luminal and lateral membranes, only about 15 times, exhibiting a geometric symmetry of luminal and lateral cell walls. In addition, a positive correlation between the surface density of inner mitochondrial membranes and basal-lateral cell membranes has been demonstrated (Pfaller and Rittinger, 1977). The significance in reduction of reabsorptive capacity is not known. It has often been pointed out that the proximal tubule, while losing some of its reabsorptive capacity in the straight portion, gains secretional capacity. By expanding upon previous experiments that demonstrate a high capacity for phenol red secretion (Rollhäuser, 1960), secretion of PAH in the straight portion has been shown to be about four times higher than in the convoluted portion (Tune et al., 1969). Many other pharmacologic and physiologic substances have been shown to be secreted by the straight portion (uric acid: Deetjen et al., 1978; urea: Kawamura and Kokko, 1976; Roch-Ramel et al., 1978; potassium: Jamison et al., 1976; prostaglandins: Grantham and Irish, 1978). However, a causal connection between decreased reabsorption and increased secretion remains unconfirmed.

In view of the discussion about a heterogeneity between different nephron types (superficial, midcortical, juxtamedullary), the morphologic differences between the different proximal tubules are astonishingly small. The rabbit shows no differences in total length of proximal tubules of different nephron types (Bankir and Rouffignac, 1976). As regards outer and luminal diameter, S3 of juxtamedullary nephrons is considerably larger than S3 of superficial and midcortical nephrons (Fig. 12b). These differences in luminal diameter can be supposed to correlate with differences in single nephron glomerular filtration rate (SNGFR), which is known to be a factor 1.25 greater in juxtamedullary nephrons (Horster and Thurau, 1968; Bonvalet et al., 1972; Bankir et al., 1975). These differences in luminal diameter among proximal tubules correspond to differences in the subsequent descending limbs. The luminal diameter of the longest long loops differs by up to a factor of 2 from that of short loops (see 4.2). As regards ultrastructure of proximal tubular segments there are no fundamental differences in superficial, midcortical, and juxtamedullary nephrons; the proximal tubule cells of different nephron types can hardly be distinguished. Only minor differences exist, e.g., S1 cells of juxtamedullary nephrons already contain numerous large peroxisomes and a high amount of rough endoplasmatic reticulum.

The similarity between proximal tubular segments in different nephrons suggests that the alterations along a proximal tubule are at least partly dependent on the localization of the different segments and thus, on different histotopographic relationships. S1 lies exclusively in the cortical labyrinth, S2 at the borders of the labyrinth and in the upper parts of the medullary rays (those of juxtamedullary nephrons are shorter!). S3 lies mainly in the outer stripe of the outer medulla. Thus, S1 occupies the most

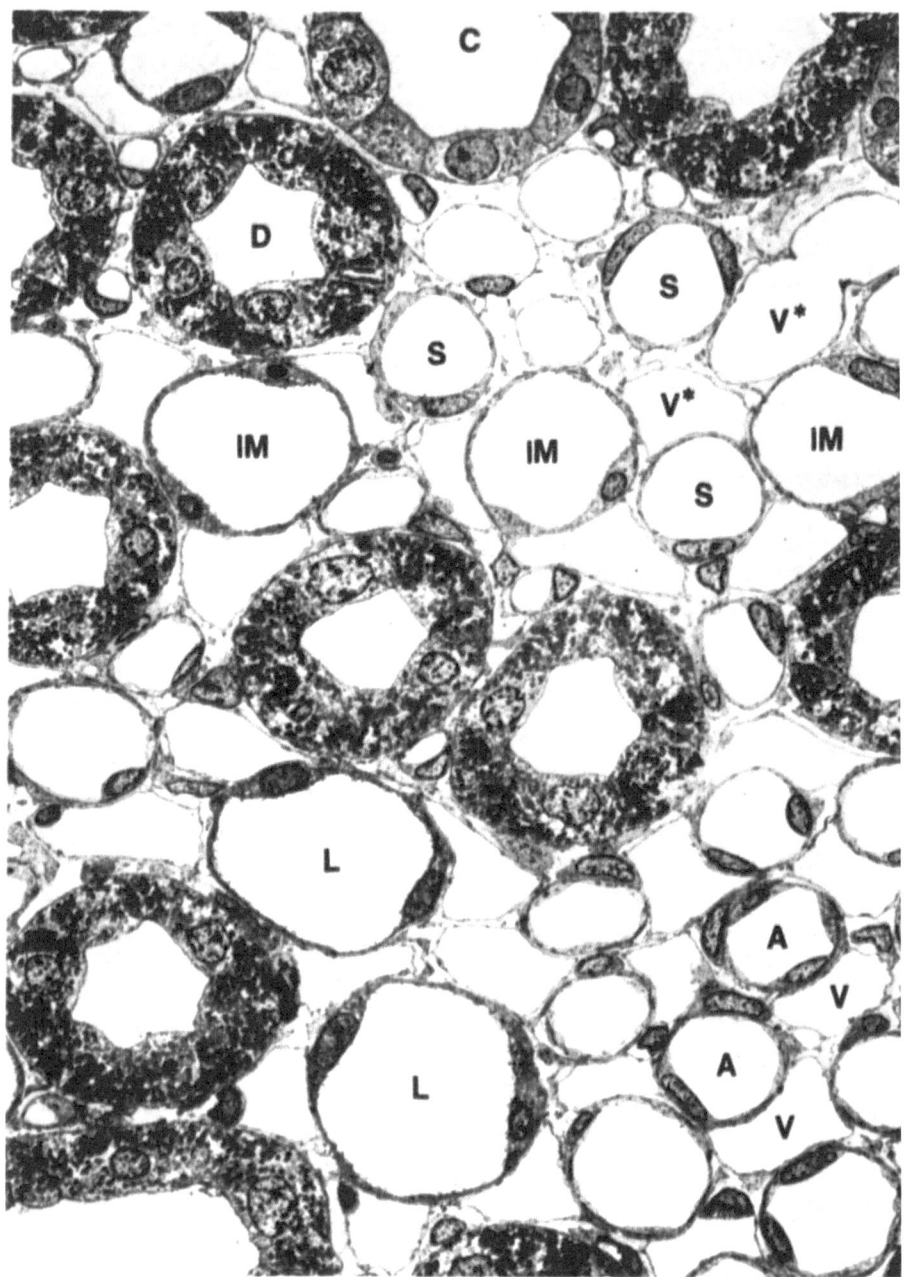

Fig. 20. Cross section through the upper part of the inner stripe. The vascular bundle only contains arterial *(A)* and venous *(V)* vasa recta. The large thin limbs *(L)* lie near the vascular bundle. The intermediate *(IM)* and small *(S)* thin limb profiles are situated distant from the bundle frequently associated with venous vasa recta *(V*)*, which ascend independently from the bundles. *D* = straight distal tubule; *C* = collecting duct. x ~900

heavily capillarized region, intermingled only with a small number of other tubules (convoluted distal tubules, connecting tubules). S2 within the medullary rays shares this region with an equal number of straight distal tubules and the corresponding collecting ducts surrounded by the less dense medullary ray capillary plexus. S3 within the outer stripe receives blood supply predominantly from ascending venous vasa recta. Thus, striking parallels are shown: S3 has a secretory function for various substances; these secreted substances must originate predominantly from the medullary venous blood; i.e., the secreted substances are all able to recycle within the renal medulla (see 3.7.4).

4.2 Thin Limb Ultrastructure

(In cooperation with J. M. Barrett, Medical College of Georgia, Augusta, USA)

A cross section through the upper part of the inner stripe (Figs. 10 c, 13 b and 20) shows that thin descending limbs of the loops of Henle are not a homogeneous population, but that they differ markedly in luminal diameter and epithelial thickness. Small diameter cross sectional profiles of thin limbs, lined by a flat epithelium with a smooth luminal surface, alternate with those of large (up to twice the size of small thin limbs) and intermediate diameters lined by a thicker epithelium bearing microvilli on the luminal surface. Even if a precise identification of the cross sectional profiles is difficult on the light microscopic level, it may be deduced from single nephron tracings (see 3.4.3) that thin limbs with bigger lumina belong to long loops, while those with small lumina belong to short loops.

These deductions are in accordance with our findings on the transition of proximal tubules to thin descending limbs at the border between outer and inner stripes (Fig. 21). The proximal tubules of juxtamedullary nephrons (which are definitely long looped), descending in the vicinity of the vascular bundles, pass over into thin limbs with large lumina. On the other hand, straight proximal tubules coming down from the medullary rays (and thus probably short looped) transform into thin limbs with small lumina. In addition, it can be demonstrated that thin limbs with a small lumen pass over, in the lowermost part of the inner stripe, directly into the thick-wall straight distal tubules, thus forming a short loop (Fig. 22). On the other hand, thin limbs with large lumina traverse the border between outer and inner zones to form a long loop (Figs. 22 and 28). Thus, it is justified to conclude that thin descending limbs with small lumina form short loops whereas those with larger luminal diameters form long loops. The largest cross-sectional profiles among the latter group are believed to belong to the longest long loops, whereas those of intermediate sizes belong to long loops of intermediate actual length.

The thin limbs of short loops of Henle. In short loops only a descending thin limb is present. In rabbit short loops, the transition from the thin-wall limb to the thick-wall straight distal tubule may occur shortly before the bend of the loop, at the very bend, or shortly after the bend within the ascending limb. In the latter case there is a very short ascending thin portion, which does not, however, differ in its ultrastructure from the descending portion; therefore, it is unnecessary to consider it separately. The epithelial transition is always abrupt without an interposed intermediate cell (Fig. 22).

The thin descending limbs of short loops (Fig. 22, 23 c and 29 a) are established by

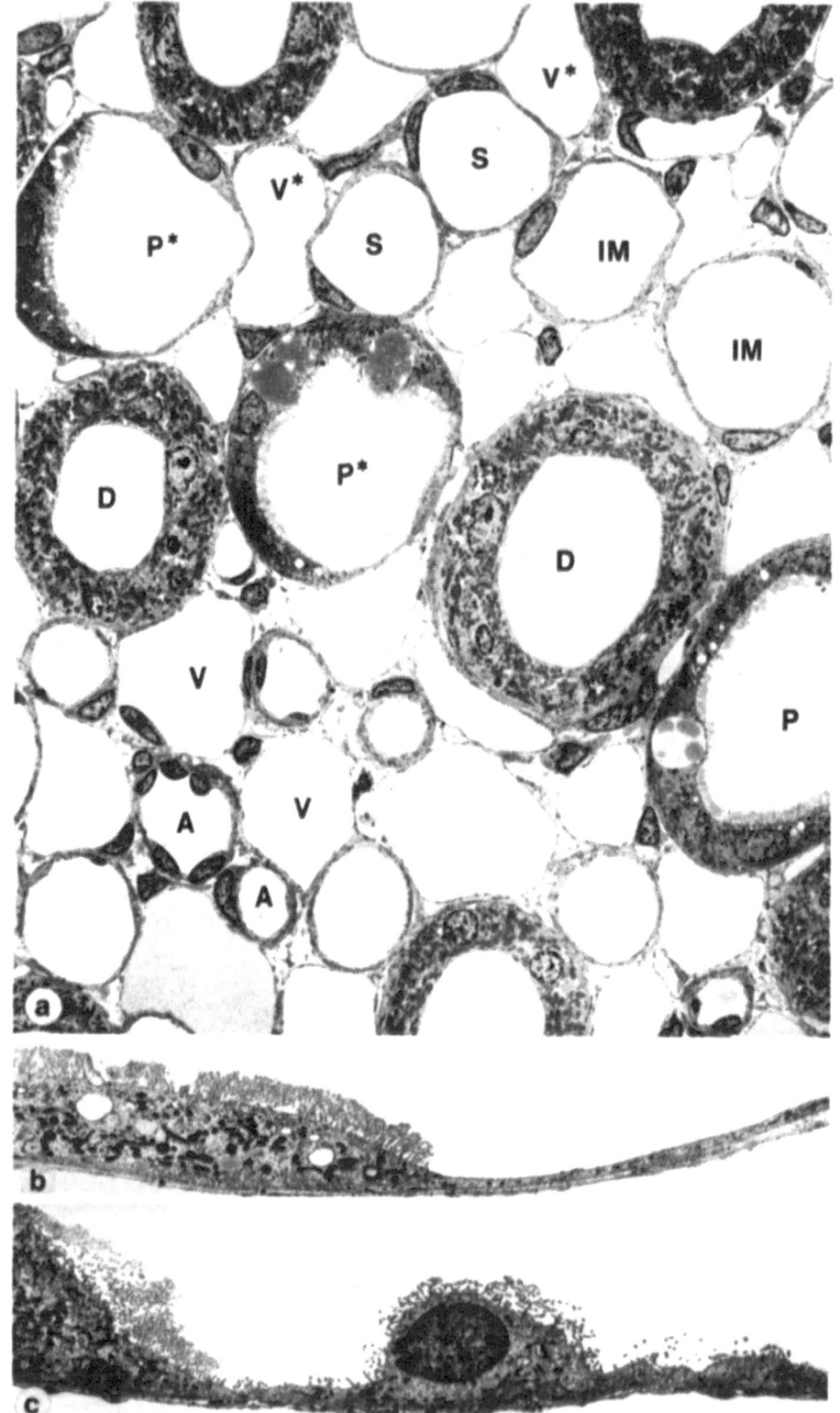

Fig. 21

52

a flat (maximal cellular height in non-nuclear regions, 0.3 μm), very simply organized epithelium. All usual cell organelles (predominantly located in the perinuclear region) are sparse and the luminal and basal cell surfaces are almost completely smooth. Basally the epithelium is wrapped by a thin, single-layer uninterrupted basal lamina.

The epithelial cells do not interdigitate; in a cross section portions of only two or three cells are encountered. The lateral cell borders of two adjacent cells may either overlap, interlock simply, or abut bluntly. The cells are joined together by a tight junctional belt (zonula occludens) of intermediate apical-basal depth (119.2 ± 42.5nm, n = 14); the tight junctions are often accompanied by a desmosome (Fig. 23 $c_1 - c_2$).

The thin limbs of long loops. In long loops a thin descending limb and a thin ascending limb are present. The thin descending limb may be subdivided into an outer medullary and an inner medullary portion. This division does not fully correspond with an ultrastructural division; the epithelium of the outer medullary portion gradually transforms into the inner medullary descending epithelium. The latter passes over, immediately or some short distance before the bend, into the ascending-type epithelium, which in turn changes abruptly at the border of inner and outer zone to the thick-wall epithelium of the distal tubule.

In the inner stripe of the outer medulla the limbs of intermediate and large cross-sectional sizes (see above) have been found to belong to long loops. Apart from differences in epithelial thickness the ultrastructural organization of large and intermediate thin limb profiles is equal.

The epithelium of the outer medullary portion of the thin descending limbs of long loops (Figs. 23 a and b) varies in its cellular thickness between 0,3 and 0.8 μm. Even in a given cross section the epithelial thickness varies since parts of the cells (in addition to the perinuclear parts) may bulge into the lumen. The cytoplasm is stuffed with a considerable amount of small mitochondria distributed throughout the cell. It contains numerous polysomes; Golgi profiles as well as lysosomal vesicles are frequently encountered. All other organelles are sparse (Fig. 29 b–d).

The apical surface of the epithelium is typically equipped with numerous stubby microvilli. The basal part of the epithelium is characterized by an intricate basal labyrinth established by the interdigitation of small irregular cellular processes all of which, in a certain area, originate from the same cell (Fig. 23 a_1). Basal to the labyrinth, the epithelium is wrapped by a thin, single-layer, and uninterrupted basal lamina.

The epithelial cells do not interdigitate with each other. In contrast to short loops, however, a cross section may be composed of several cells. The thickest profiles may contain slices of up to eight cells. This may be interpreted as a moderate degree of a first-order interdigitation. There are, however, no secondary processes as in the ascending type epithelium, which would establish it as an interdigitated epithelium. The lateral cell borders may be simply apposed to each other, may overlap, or may be

---

Fig. 21 a–c. Transition between outer and inner stripe. *a.* The proximal tubules *(P)* of juxta-medullary nephrons are situated around the vascular bundles. Two of them *(P*)* pass over into the thin descending limbs with epithelium exhibiting the characteristics of the large thin limb type. The transitions to thin limbs of intermediate *(IM)* and small *(S)* size have already occurred somewhat earlier. *D* = straight distal tubule; *A* = arterial vasa recta; *V* = venous vasa recta; *V** = venous vasa recta ascending independently from the bundles; x ~1000. *b.* Transition of a proximal tubule into a thin limb of a small type, i. e., of a short loop of Henle; x ~2700. *c.* Transition of a proximal tubule into a thin limb of the large type, i. e., of the long loop; x ~2750

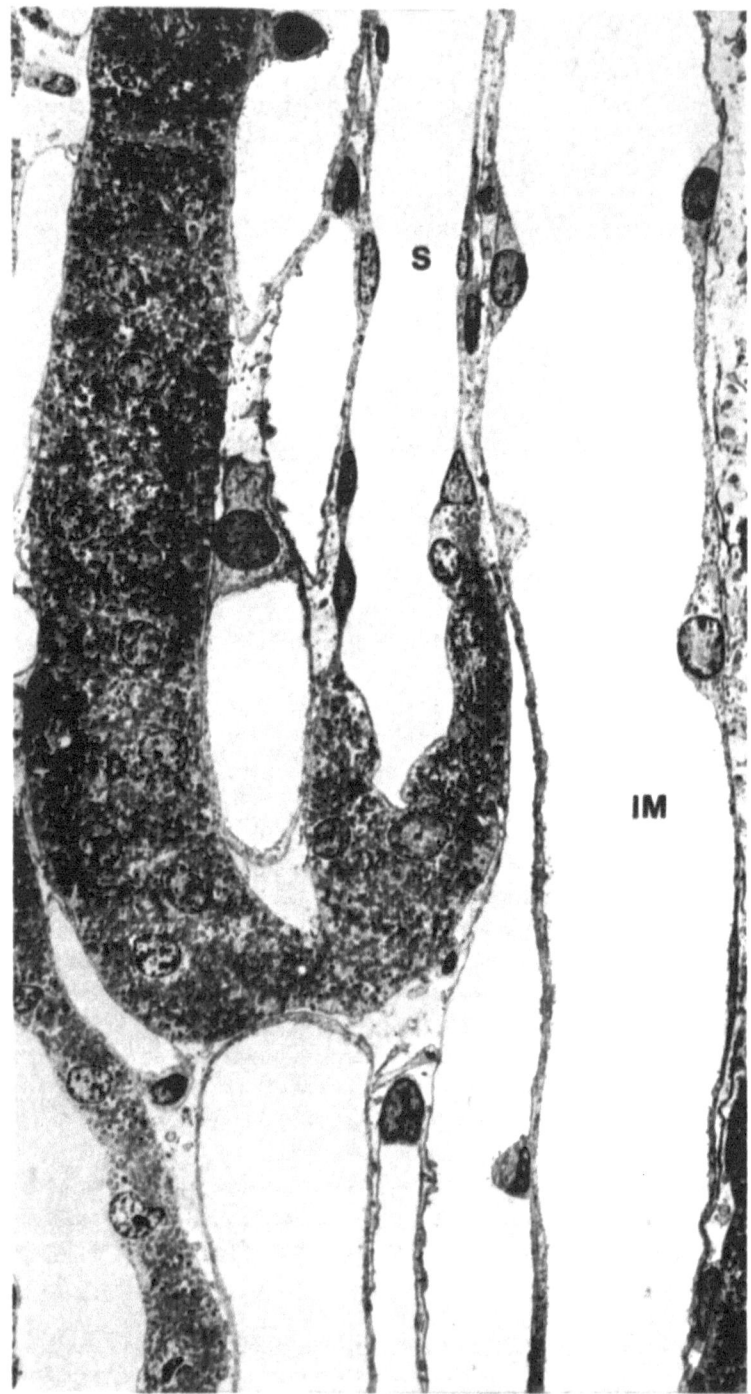

Fig. 22. Bend of a short loop of Henle located in the lowermost part of the inner stripe. The thin descending limb is classified according to its diameter and the character of its epithelium under the small type (S). The bend is already established by the thick epithelium of the straight distal tubule (D). Another thin descending limb belonging to the intermediate type (IM) descends into the inner zone. x ~1000

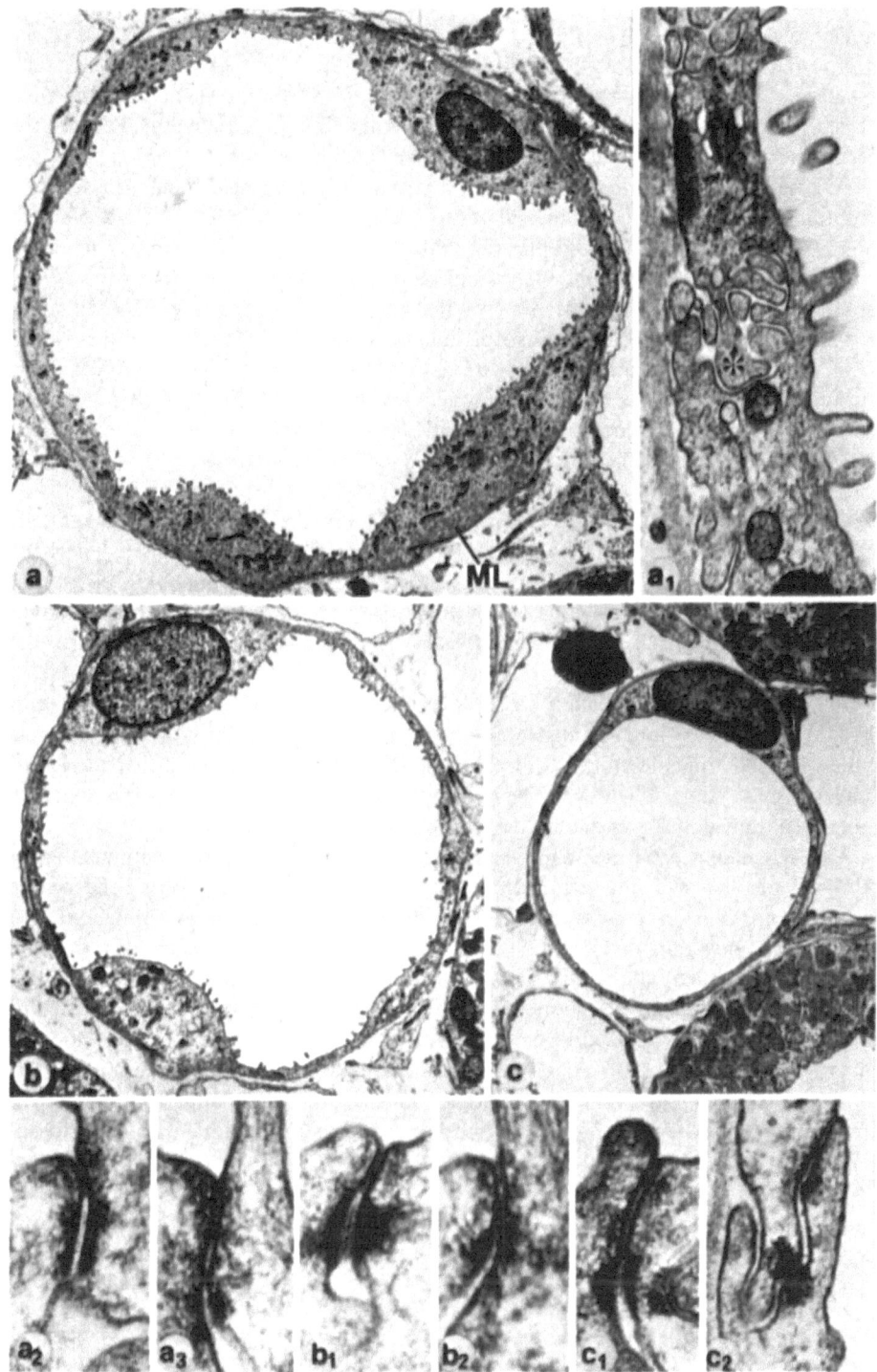

Fig. 23 a-c. Thin descending limbs in the inner stripe. *a*. Large thin limb cross-sectional profile; epithelial cells bulge into the lumen and their apical surface is covered by numerous microvilli. At their base an obvious membraneous labyrinth *(ML)* is established as shown in *(a₁)* by true in-

moderately interlocked. The tight junctions (Figs. 23 a_{2-3} and 23 b_{1-2}) are of an intermediate apical-basal depth (in loop limbs of intermediate size: 96.5 ± 34.0 nm, n = 25; in loop limbs of large size: 92.6 ± 30.0 nm, n = 28) not convincingly different from those in the short descending limbs; they may also be accompanied (probably less frequently than in the shorter loops) by desmosomes.

Descending into the inner zone, the limb epithelium changes gradually in its ultrastructural appearance, becoming still more simple (Figs. 24 and 25 a). The apical microvilli have almost fully disappeared and the basal labyrinth is strongly reduced. Only remnants of the latter may be found more frequently in the upper parts of the inner zone than at deeper levels. The cellular height, which is lower in the beginning of the inner zone than in the inner stripe, increases again towards the tip of the papilla (Figs. 11 c and 27). Also the number of cell organelles (mitochondria, polysomes) is smaller in the upper parts of the inner zone than in the inner stripe. Thicker epithelial cells within the papilla once again contain somewhat more mitochondria and, in addition, a considerable amount of vesicles of lysosomal appearance.

Throughout the inner zone, a cross-sectional profile of a descending limb (Fig. 25 a) contains slices of four to five cells. The cell borders of the majority of adjacent cells are simply apposed to each other and sometimes moderately interlocked. The tight junctions (Fig. 25 a_{1-2}) tend to increase in their apical basal depth (121.1 ± 67.4 nm, n = 17) within the inner zone; a clear-cut difference compared to those in the inner stripe is not evident. Desmosomes, however, are very seldom encountered in the inner zone.

The epithelium is underlaid by a basal lamina, which is thin with a single layer in the beginning of the inner zone; towards the papilla it tends to split into several layers (all of original thickness), with the central layers breaking down into particles of irregular size. Thus, the total basal lamina thickens considerably to form a belt, a process continued in the ascending thin limb.

The descending type epithelium passes over into the ascending type immediately at the bend or just a few cells upstream (Figs. 26 and 27). The very bend of the loops has mostly been found to be lined already by the ascending type epithelium.

The ascending thin limbs (Figs. 25 b, 28 and 29 e) are established by a thin but heavily interdigitated epithelium (cellular height in the non-nuclear regions, ~0.5 μm; within the papilla, as in the descending limbs, the limb epithelium is thicker, reaching a total thickness of ~0.8 μm). In a cross section up to 50 individual cell processes may be encountered. They are joined together by very shallow, tight junctions (32.7 ± 13.0 nm, n = 26; Fig. 25 $b_{1,2}$) consisting, as far as this information may be derived from routine thin sections, of merely one or two, maximally three ridges; additional

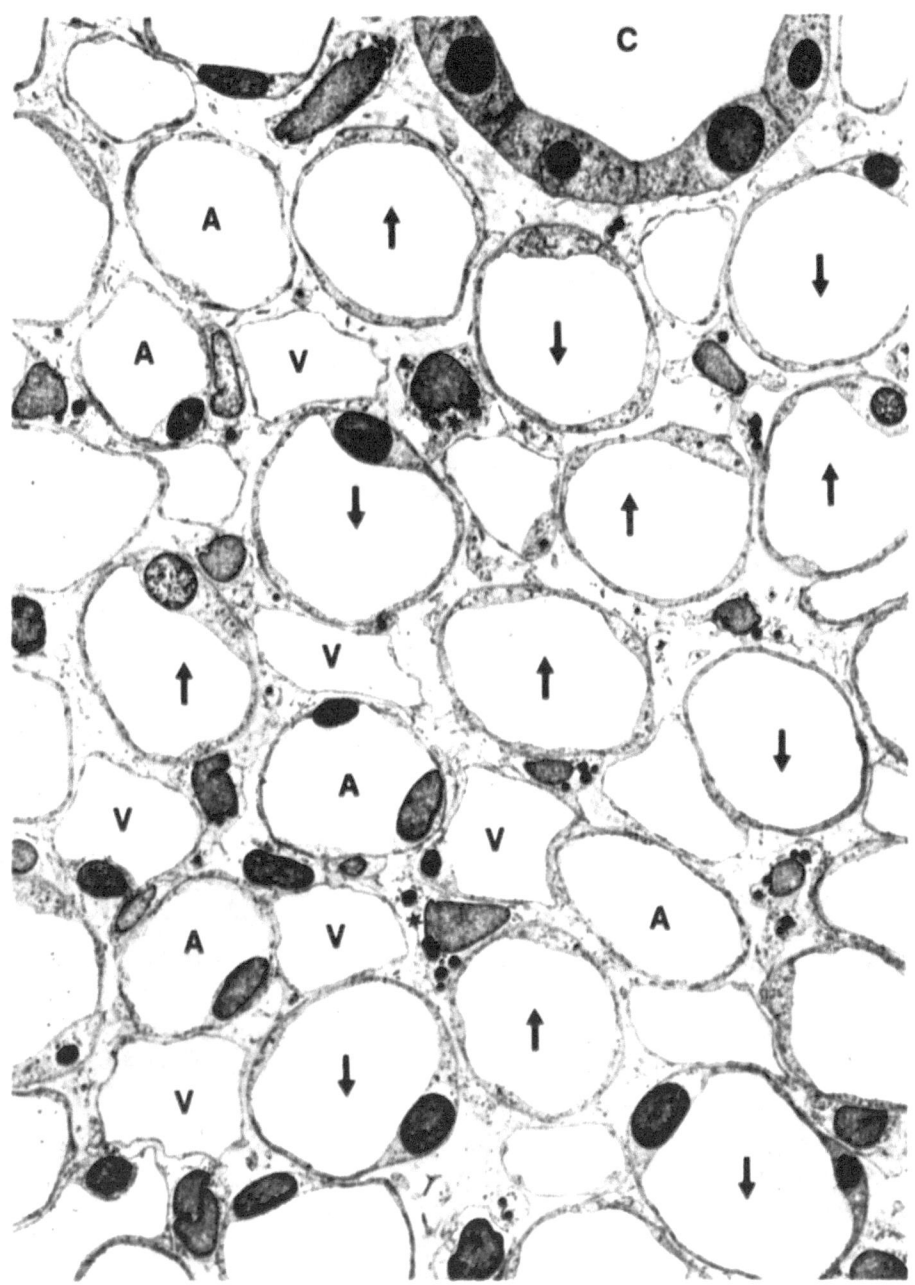

Fig. 24. Cross section through the upper part of the inner zone. The vascular bundle is not clearly delimited. Arterial *(A)* and venous *(V)* vasa recta together with thin descending *(↓)* and ascending *(↑)* thin limbs (which at this magnification cannot be clearly distinguished; the marcations have been controlled at higher magnifications) are embedded in a common interstitial space. Capillaries cannot be distinguished from venous vasa recta. *C* = collecting duct; *asterisk* = interstitial cell. x ~1000

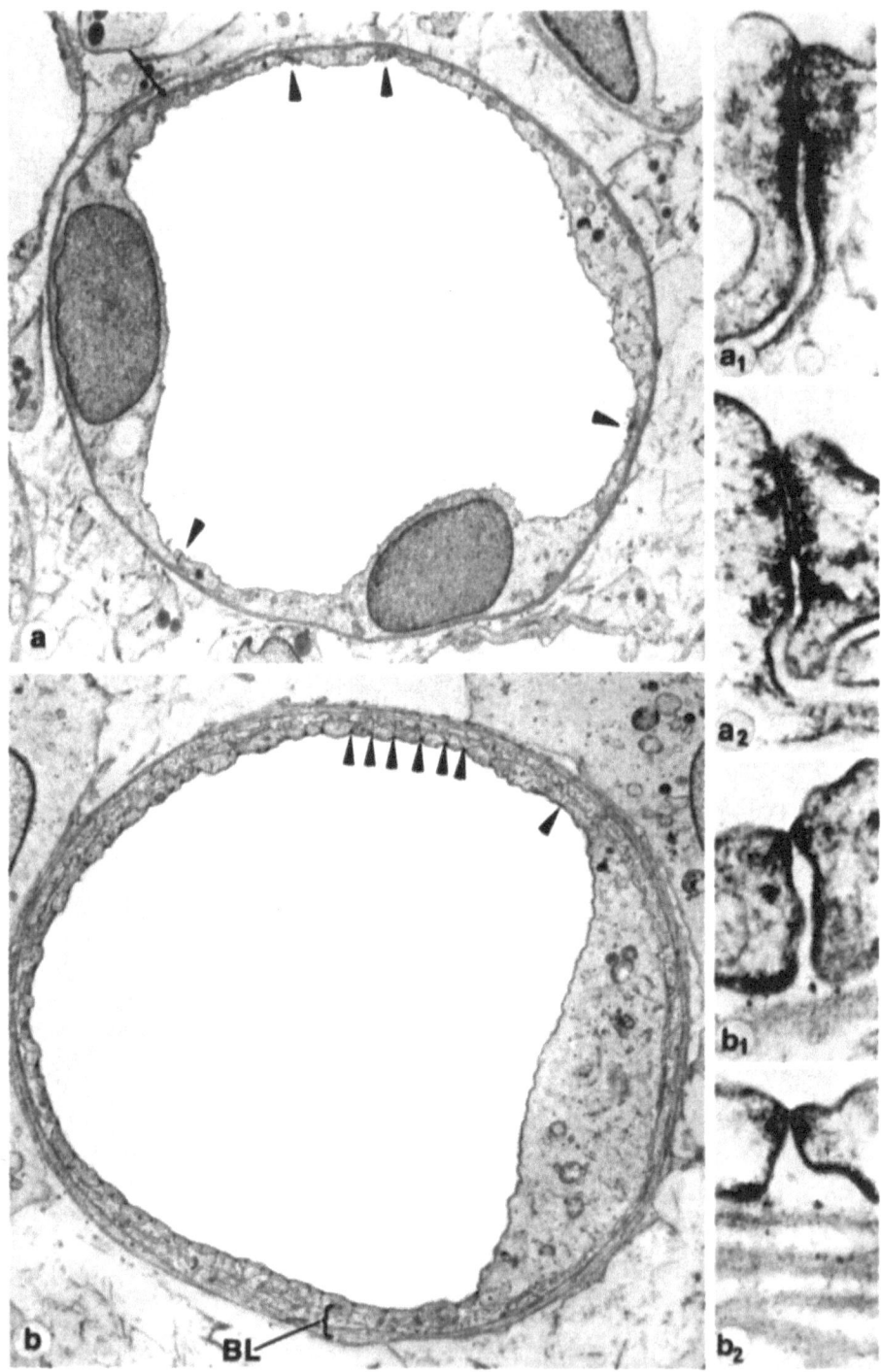

Fig. 25, a and b (Caption, see p. 60)

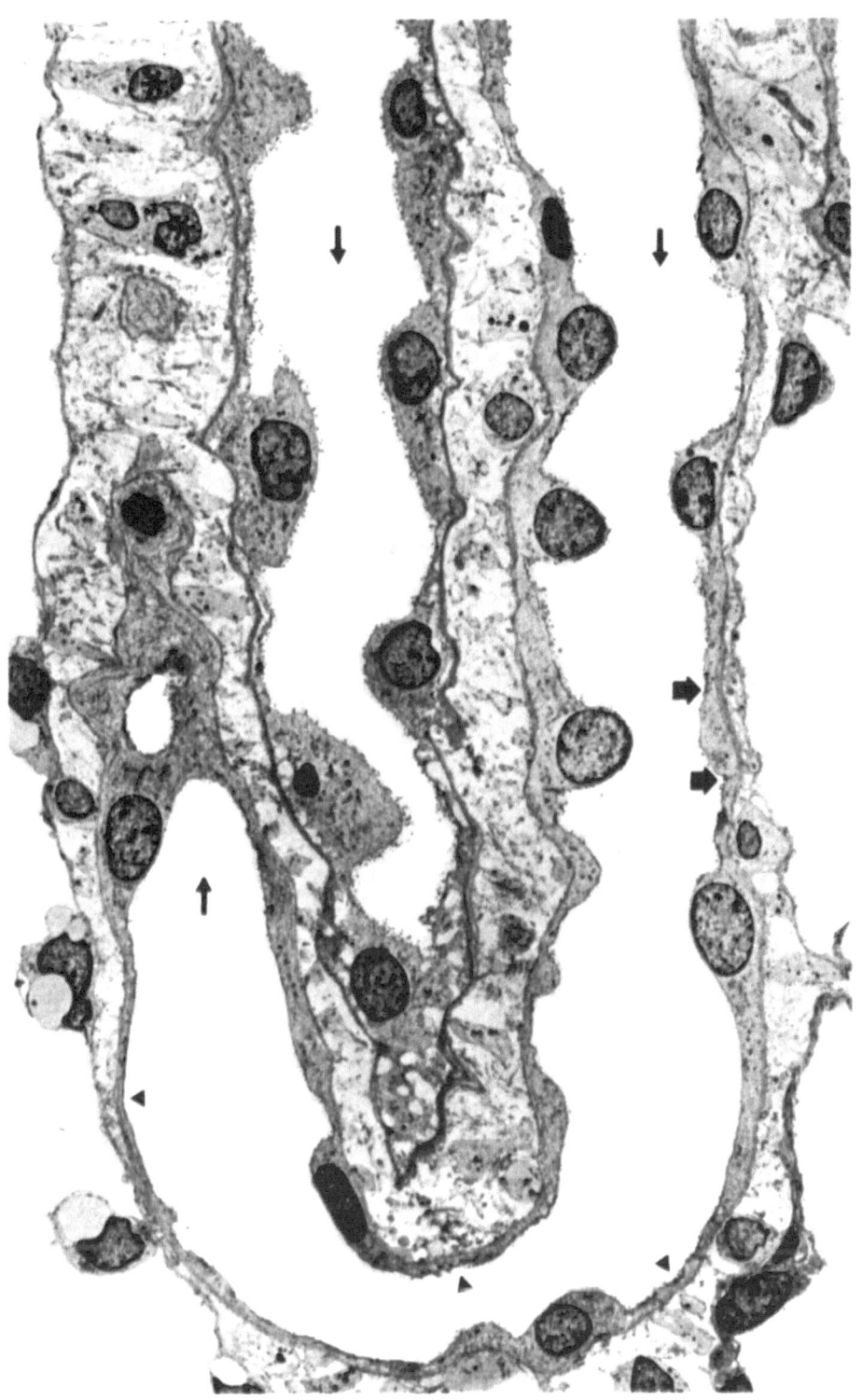

Fig. 26 (Caption, see p. 60)

desmosomes have never been found. Thus, the most characteristic feature of this epithelium is the extraordinarily high amount of shallow paracellular pathways. Beneath the tight junctional barrier the cell processes usually narrow leaving between them funnel-shape clefts, which open toward their base.

The basal surface is irregular because the interdigitating processes vary in their thickness. The irregular base of the epithelium is smoothed by the basal lamina, which consists of several layers (Figs. 25 b and 27). The outer layer(s) is (are) uninterrupted, whereas the more central layers are broken into fragments of different size. They also fill the spaces between the cellular processes, often reaching up to the tight junctional barrier.

The cytoplasm regularly contains a moderate amount of small mitochondria. In addition, in the papilla (Fig. 27) as in the descending thin limbs, numerous lysosomal vesicles may be encountered. All other cell organelles are sparse.

Considerable differences become obvious when comparing cross-sectional overviews of the upper part of the inner zone with those of the papilla (Figs. 11 a versus 11 c and 24 and 26 versus 27). In addition to a drastic increase in cellular height of the collecting duct cells (see. 4.4.2), the epithelium of both loop limbs is thicker in the papilla. At both levels the loop limbs are randomly distributed. Together with the vascular components (arterial and venous vasa recta and capillaries) and collecting ducts, the loop limbs are embedded into a well-developed common interstitial space, the amount of which increases towards the papilla. It contains the typical interstitial cells (Fig. 27), which also increase in number towards the tip of the papilla. These stellate-shape cells with long slender processes contain numerous lipid droplets, which also increase in number toward the papilla.

Fig. 25, a and b. Upper half of the inner zone. *a*. The cross-sectional profile of the thin descending limb is established by only five epithelial cells (the junctions are marked by *arrow heads*). Remnants of a membraneous labyrinth *(arrow)* are present. The basal lamina is unilayered; only at some sites can a splitting be observed. The tight junctions *(a₁ and a₂)* of the thin descending limbs are of intermediate apical- basal depth. *b*. The thin ascending limb is established by a heavily interdigitated epithelium; 46 cell processes (junctions are partly marked by *arrow heads*) are encountered in this cross-sectional profile. The basal lamina *(BL)* is split into several layers. The tight junctions of the thin ascending limbs *(b₁ and b₂)* are extremely shallow. *(a and b)* x ~4000; *(a₁, a₂, b₁, b₂)* x ~88500

Fig. 26. Bend of a long loop of Henle in the upper part of the inner zone. The transition from the descending type epithelium to the ascending type occurs within the descending (↓) loop limb: the first shallow tight junctions (characteristic for the ascending epithelium) are marked by *thick arrows*. Shallow tight junctions at the bend and in the ascending (↑) loop limb are marked by *arrow heads*. Another descending loop limb (↓) is present. Note the transversely oriented interstitial cells *(asterisks)*. x ~1400

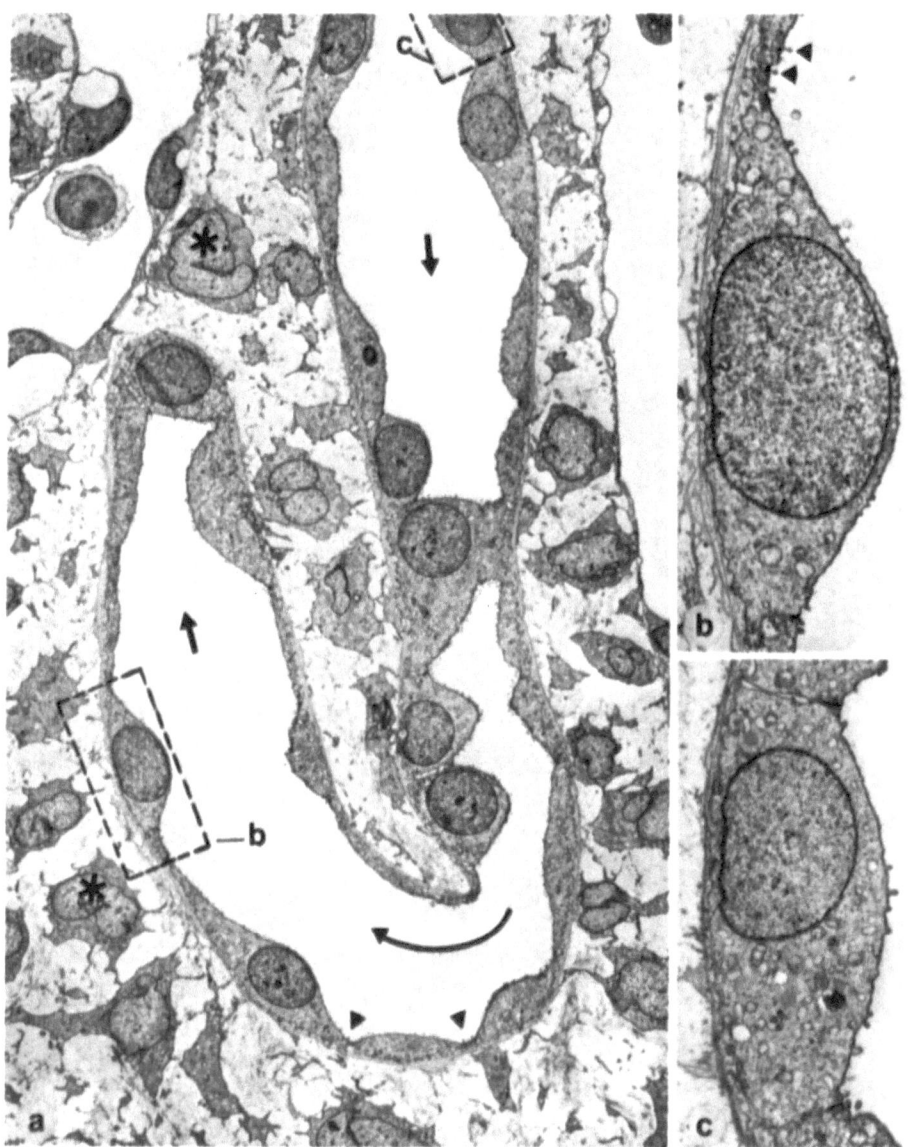

Fig. 27 (a–c). Bend of a long loop of Henle from the papilla. Typical signs for the ascending type epithelium (interdigitating cell processes) are first observed at the very bend of the loop *(arrow heads)*. Accordingly the transition from the descending to the ascending type epithelium, in this case, fully corresponds with the point of looping back. *b.* A higher magnification of the ascending epithelium (marked in *a*). *c.* A higher magnification of the descending epithelium (again marked in *a*). Note the well-developed interstitium with the numerous interstitial cells *(asterisk). a.* x ~1350; *(b)* and *(c)* x ~4500

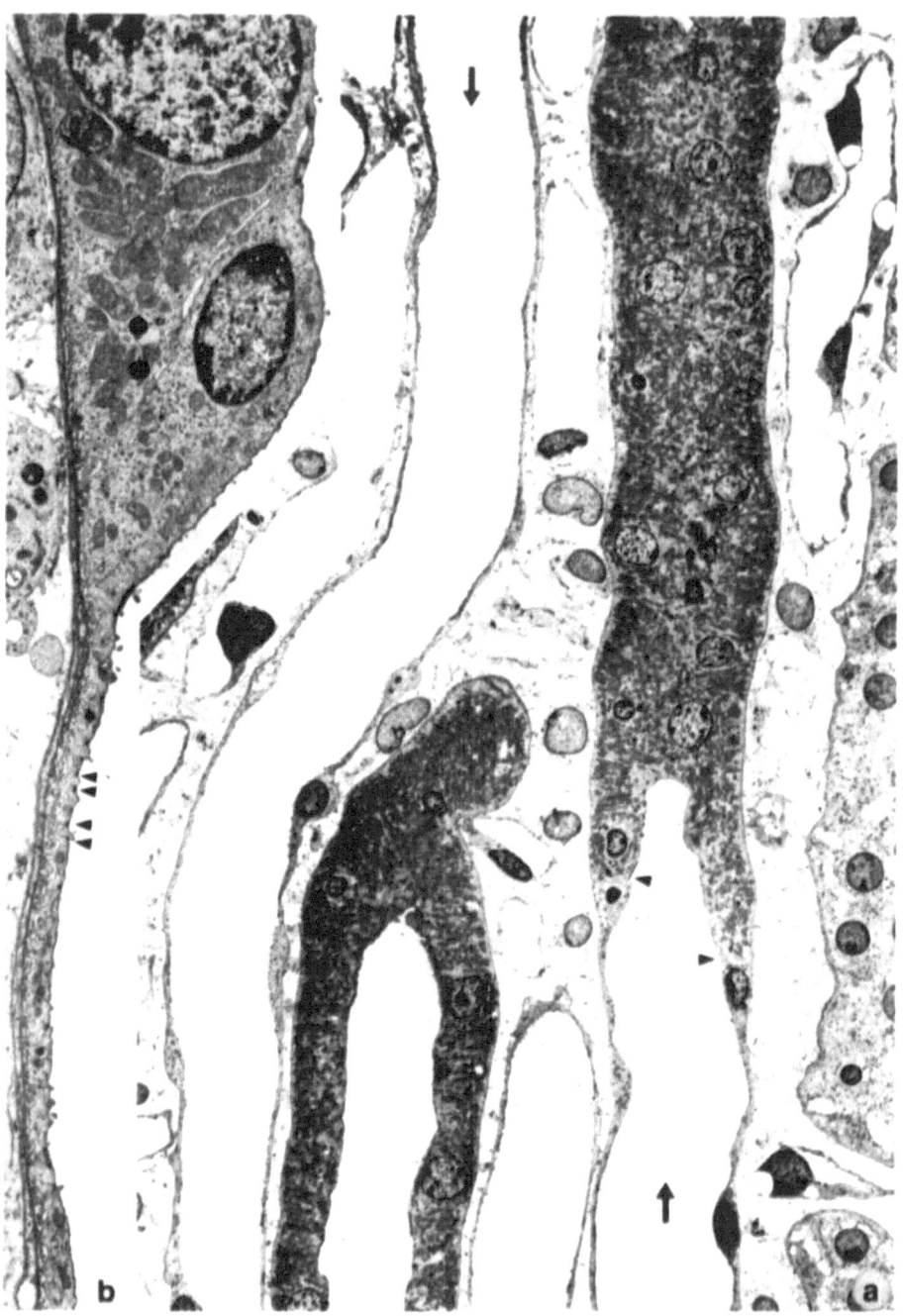

Fig. 28, a and b. Border between inner stripe and inner zone. *a.* An ascending thin limb *(↑)* of a long loop of Henle comes from the inner zone and makes its transition *(arrow head)* to the straight distal tubule *(D).* Additionally a descending limb *(↓)* of the large / intermediate type traverses the border between outer and inner zones. *b.* A higher magnification of a transition from an ascending thin limb to the thick epithelium of the straight distal tubule. The ascending limb epithelium is typically characterized by interdigitating cell processes connected by shallow junctions *(arrow heads).* *(a)* x ~820; *(b)* x ~4400

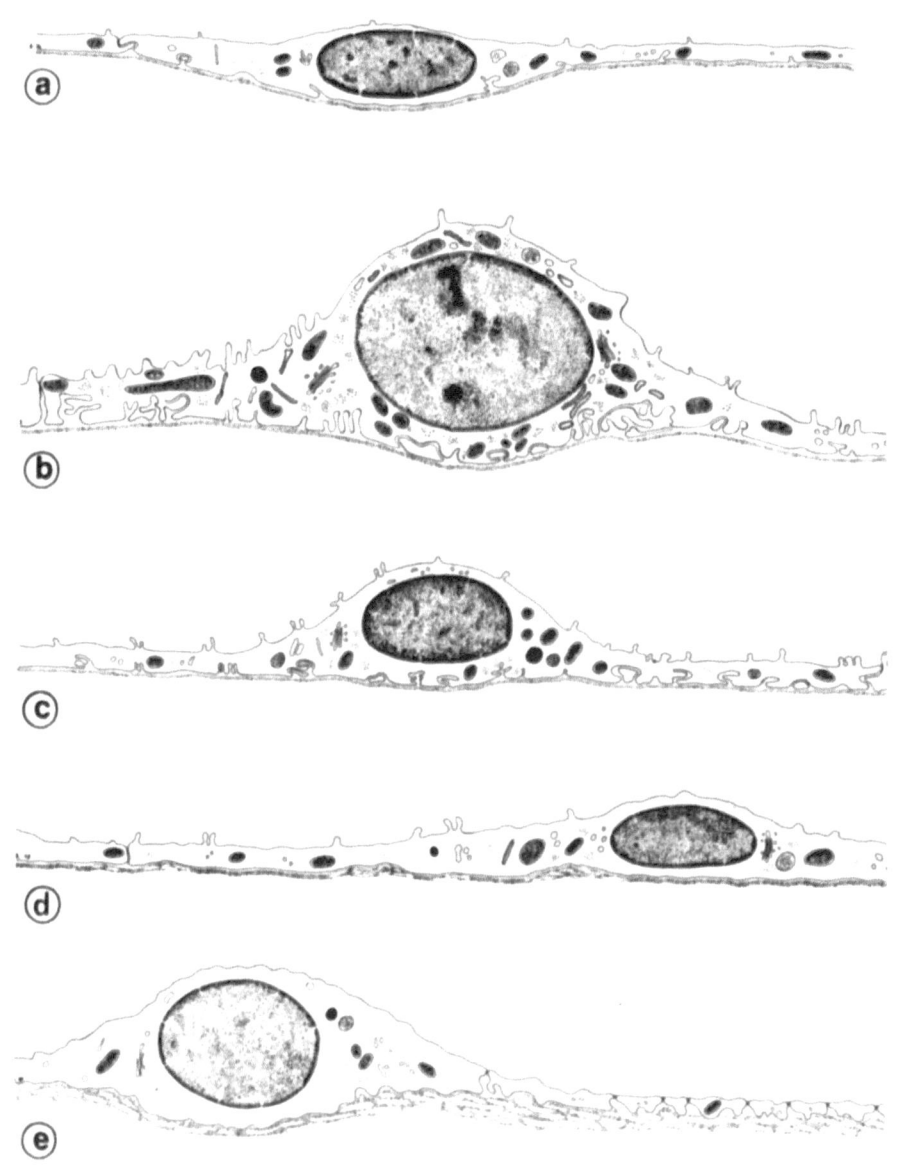

Fig. 29, a-e. Thin limb cells: *a.* Descending thin limb of a short loop; *(b)* outer medullary portion (large type) of a long descending limb; *(c)* outer medullary portion (intermediate type) of a long descending limb; *(d)* inner medullary part of a long descending limb; *(c)* ascending thin limb. x ~4500

4.2.1 Comparative and Functional Aspects

Since it has become known that the thin limbs of short and long loops in rats differ markedly in their ultrastructural organization (Kriz et al., 1972 b), thin limb epithelia have been systematically investigated in the rat (Schwartz and Venkatachalam, 1974), mouse (Dieterich et al., 1975), Octodon degus (Barrett and Majack, 1977), and Psammomys (Barrett et al., 1978 a, b); some preliminary data are known about the loops of the Syrian hamster and cat (Kriz et al., 1978).

Before beginning a comparative discussion, a weak point of all these investigations has to be clarified. The data about junctional complexes are all obtained from routine thin sections. A fully satisfactory evaluation of the junctions, however, would require complementary data from freeze-fracture replicas. These data are, at present, available only from two different segments of the rat thin limbs, i. e., from the outer medullary portion of the descending limb of long loops and from the thin ascending limbs (Humbert et al., 1975; Pricam et al., 1974). In both thin limb portions the tight junctions are equally developed and consist of but one to two continuous junctional ridges. Nevertheless these data are in good accord with investigations in thin sections in which the tight junctions of the corresponding thin limb segments of the rat at both sites have been found to be of a very short apical-basal depth (Schwartz and Venkatachalam, 1974). Thus, even if a definite description of the tight junctions (number and kind of ridges) cannot be given from thin sections, a relative judgement of the apical-basal depth is fully possible. In the present investigation the apical-basal depth has been measured at those sites where the tight junctional belt could be supposed to be cut in a right angle. The values obtained might be questioned as in general being too high. But taken as relative values they clearly demonstrate that in the rabbit (in contrast to the rat) only the thin ascending limbs are connected by shallow tight junctions, whereas in all descending thin limb segments they are of an intermediate apical-basal depth. A separation of the junctional membranes over the full depth of the junction, as has been described to occur in the descending thin limbs (Darton, 1969), has never been observed.

Comparing data from the rabbit with those of all other investigated species, some statements of more general importance are possible. In all investigated species the descending thin limbs of short loops are organized similarly; they are constituted by the simplest type of thin limb epithelia: flat, noninterdigitating epithelial cells connected by tight junctions of an intermediate apical-basal depth. They are very uniform in luminal diameter and cellular thickness. This fact gains interest since the thin descending limbs of short loops in different species have different histotopographic relationships. In the rat, mouse, Octodon degus, and Psammomys they are integrated into the vascular bundles running side by side with the venous vasa recta. In the rabbit and Syrian hamster, they are randomly distributed in the interbundle regions. Their similarity in ultrastructural organization suggests similar permeability and transport properties and hence a similar function at both sites, i. e., within the bundles in Psammomys or outside the bundles in the rabbit. This favors the interpretation that the integration of the short thin limbs into the vascular bundles in some rodent species with high urine concentration capability must be regarded as an improvement of the functional possibilities of the short descending limbs (Kriz et al., 1976). A further, interesting point is that in the cat, with a kidney containing exclusively long loops, many of the descending thin limbs resemble in their ultrastructural organization the descending thin limbs of short loops in the other investigated species.

At their beginning on the inner stripe the descending limbs of long loops vary considerably in luminal diameter and cellular thickness. The long loops of Henle being very different in their actual length lead us to the assumption that luminal diameter and cellular thickness in the inner stripe are correlated with the actual length of each individual loop. This interpretation fits well with the finding that in Psammomys, where the long loops all descend for a considerable distance down into the inner medulla, the scatter of the luminal diameters is less prominent than, for instance, in the mouse. One might speculate that there is a correlation of the larger lumina to the single nephron glomerular filtration rate (SNGFR), which is greater in juxtamedullary nephrons than in others (Rouffignac et al., 1970; Wright and Giebisch, 1972). The functional relevance of this, however, is unknown.

Regarding the ultrastructural organization of the descending thin limbs of long loops we have to realize, first, that there are considerable species differences and, secondly, that the epithelium changes its ultrastructural character passing down from the inner stripe into the inner zone. In the rat (Schwartz and Venkatachalam, 1974), mouse (Dieterich et al., 1975), Octodon degus (Barrett and Majack, 1977), and Psammomys (Barrett et al., 1978 b) the descending thin limbs of long loops start with a very heavily interdigitated epithelium, which in the degree of interdigitation even surpasses the ascending type epithelium. The interdigitated processes are connected by extremely shallow tight junctions. Just this predominant characteristic of extraordinarily developed paracellular pathways is lacking in the rabbit. The descending thin limbs of long loops in rabbits start in the inner stripe with, when compared to Psammomys or the rat, a simply structured epithelium. The noninterdigitating epithelial cells are connected by junctions of an intermediate apical-basal length. In the Syrian hamster still another epithelial organization has been found in some (most probably in those of the longest long loops) of the long descending limbs: high epithelial cells that are predominantly interdigitated in their apical parts, richly stuffed with mitochondria, and equipped with numerous short microvilli at the luminal site (Kriz et al., 1978).

Thus, we are confronted with the fact that the outer medullary portions of the descending thin limbs of long loops in different species are constituted of very different epithelia and must, therefore, be expected to differ markedly in their transport and permeability properties. The controversy as to whether the concentration of tubular fluid within the descending limbs occurs predominantly by solute addition (de Rouffignac et al., 1973; Imbert and de Rouffignac, 1976) or predominantly by water extraction (Kokko, 1970, 1974) may at least partly be due to these species differences in the ultrastructural organization of their thin limb segments.

Passing down into the inner zone, the epithelium of the long descending limbs in all investigated species changes gradually. At the end of this process the epithelium of the inner medullary portions of the long descending limbs has nearly the same ultrastructural character in all animals investigated. It is simply structured and noninterdigitating with tight junctions of an intermediate apical-basal length.

In the rabbit, this simplification of the epithelium is reached already at the beginning of the inner zone. The difference between outer and inner medullary segments is not fundamental in character; it mostly concerns the intricate basal labyrinth, the amount of mitochondria, the luminal diameter and, to some extent, the cellular thickness. All these features become reduced in the inner zone, compared to the outer medullary epithelium. This gradual reduction already occurs within the inner

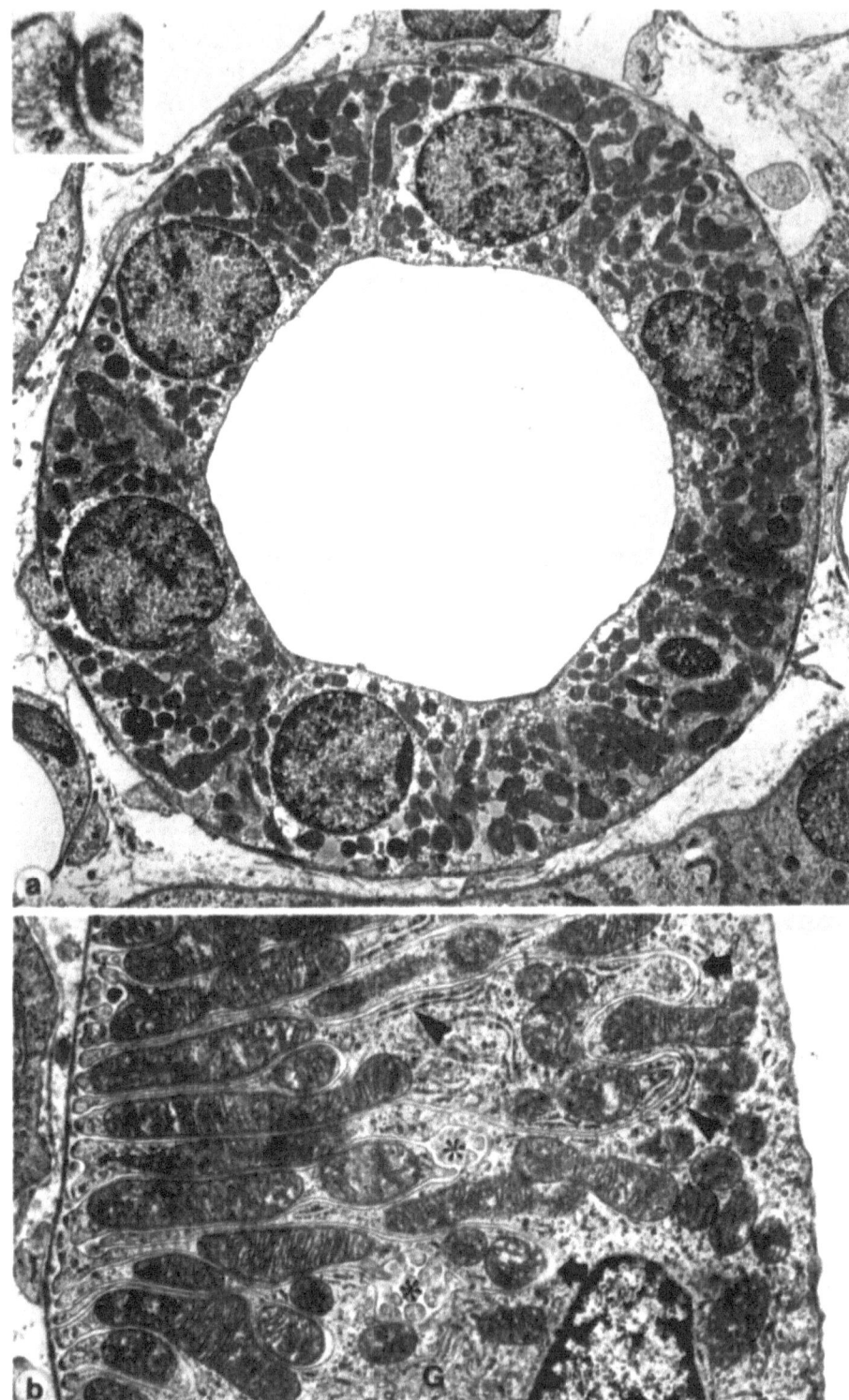

Fig. 30

stripe, that the largest cross-sectional profiles are reduced to the status of those of intermediate size. Thus, in the rabbit already at the beginning of the inner zone, the descending limbs show a homogeneous appearance with respect to luminal diameters, cell height, and their ultrastructure. In contrast, in Psammomys (Barrett et al., 1978 b) or in the rat (Schwartz and Venkatachalam, 1974) the process of reducing the complicated epithelium of the outer medulla to the simple epithelium in the inner medulla requires a longer distance and extends well beyond the border between inner and outer zones. Thus, in these species in the upper part of the inner zone a heterogeneous population of descending thin limbs is still evident which consists of thin limbs lined already by the typical simple inner-zone epithelium and others still lined by the more complex outer medullary epithelium in different stages of reduction.

The thin ascending limbs again turned out to be equally organized in all species investigated: they are established by flat but heavily interdigitated epithelial cells connected by very shallow tight junctions. Thus, the characteristic feature of this epithelium is the increase of paracellular pathways; the elongation of the zonula occludens is, again, parallelled by a reduction of its apical-basal length. Another common feature of all investigated species is that the ascending-type epithelium starts already some short distance before the actual bend. In the rat, the length of the pre-bend segment has been found to range between 56 and 133 μm (Schwart and Venkatachalam, 1974); in the rabbit a range of this distance between 0 and 140 μm has been found. The actual bend appears to be generally constituted of the ascending-type epithelium.

The uniform ultrastructural organization of the thin ascending limbs in various mammals agrees with functional data: in the rabbit (Imai and Kokko, 1974) and rat and hamster (Imai, 1977) a comparably high ion permeability (Na^+, Cl^-) has been found.

4.3 Distal Tubule Ultrastructure

Based on ultrastructural criteria, the distal tubule can be subdivided into a medullary straight part, a cortical straight part (together corresponding to the thick ascending limb of Henle's loop), and a convoluted part. The transition from the medullary to the cortical straight part is gradual. For purposes of definition, the medullary straight part lies within the outer medulla (inner and outer stripe), the cortical straight part, within the medullary rays of the cortex; it penetrates into the cortical labyrinth, abruptly passes some short distance beyond the macula densa over into the convoluted

---

Fig. 30, a and b. Distal tubule, straight medullary part. *a.* Cross-section, a short distance beyond the transition from the thin limb. The nuclei with a rather homogenous chromatin structure are situated in the center of the cells. The cells are richly stuffed with mitochondria and the apical cell membrane is smooth. x ~ 3600. Insert: tight junction in the straight medullary part of the distal tubule; x ~ 85000. *b.* Longitudinal section through a cell in the midregion of the inner stripe. The parallel arrangement of mitochondria within large interdigitating cell processes *(thick arrows)* is obvious. Basally, the interdigitations are ramified *(thick arrows)*, and in the center of the cell they are often further split up into small villi *(asterisk)*. The nucleus has shifted up to the apical cell region, and the Golgi-apparatus *(G)* lies beneath; arrow head = cisterns of rough endoplasmatic reticulum, arranged as PTS; x ~12800

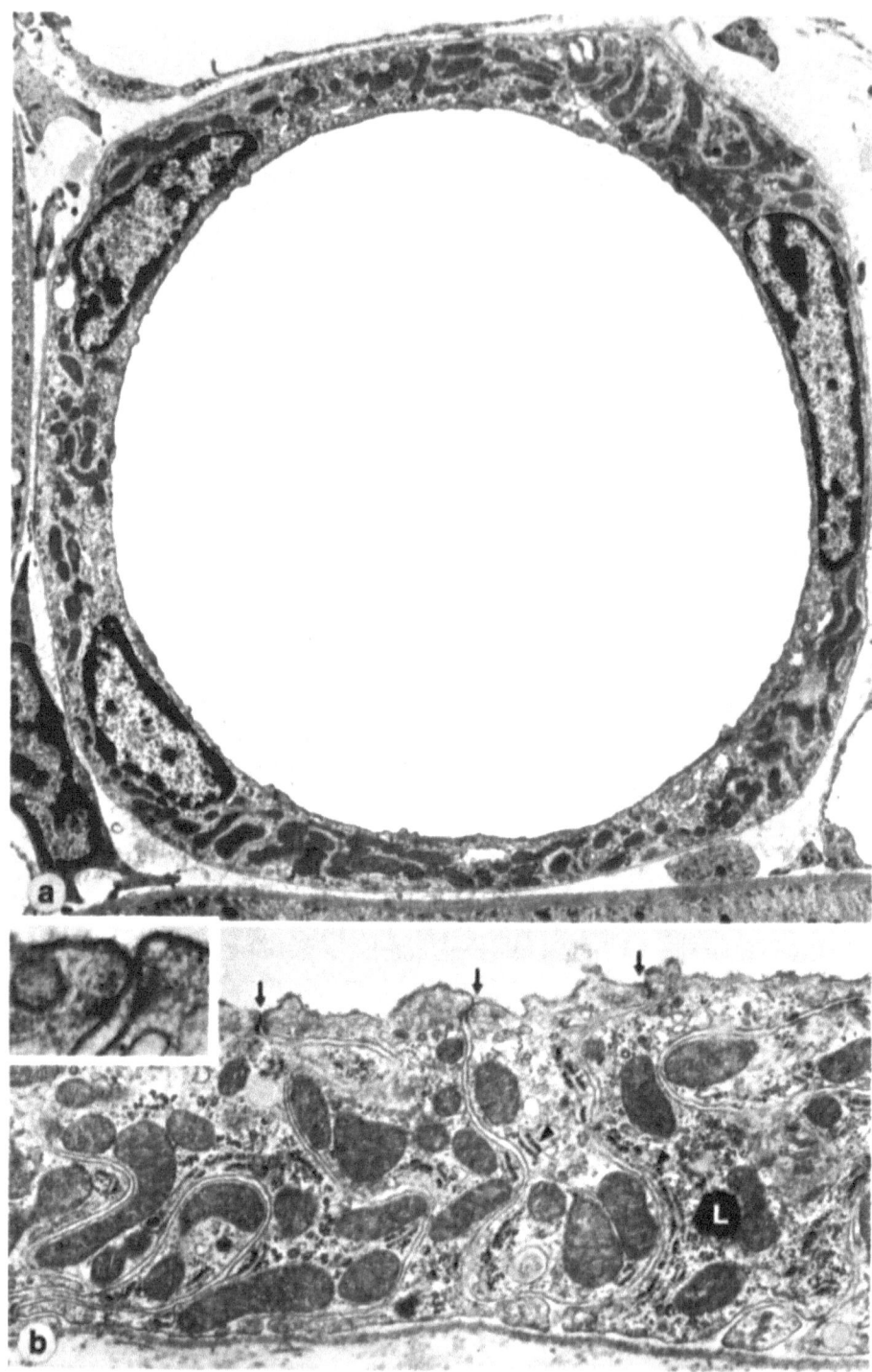

Fig. 31

part situated exclusively within the cortical labyrinth. The border between the convoluted part and the subsequent connecting tubule can also be clearly demarcated.

Medullary straight part of the distal tubule (Fig. 30). The epithelium of the medullary straight part of the distal tubule is composed of one single type of cells. The luminal quarters of the cells fit together smoothly and the basal three-quarters are extensively interdigitated. The interdigitating processes are roughly parallel in basal-apical direction and contain large rod-shaped mitochondria with a rather light mitochondrial matrix. In their basal parts the processes possess short slender ramifications (basal villi), which slide one under the other and lie as small feet on the basal lamina. They contain no mitochondria but regularly show a condensation of filaments in the basal cytoplasm. The cells are joined by shallow tight junctions (43.9 ± 23.7 nm, n = 36; derived from the entire straight part). In the rat they have been described as permeable for lanthanum (Tisher and Yarger, 1973). Gap junctions and desmosomes are absent. The slightly convex apical membrane is smooth. Cilia may be found on each cell. The big spherical nucleus (with rather homogeneous chromatin structure) is positioned centrally and occupies almost the total cell height. The Golgi apparatus is localized laterally or basally to the nucleus and is composed of small vesicles and few unexpanded cisternae. Long cisternae of rough endoplasmatic reticulum (RER) are predominantly present as paramembraneous tubular systems (PTS) following the lateral cell membrane. Polysomes are distributed all over the cell. A few small, uncoated vesicles and microtubules can be found under the apical membrane and a few lysosomal cell organelles and lipid droplets may be present in all regions of the cell. Peroxisomes have not been encountered. Cytochemically, however, a few small microperoxisomes have been found (guinea pig: Novikoff et al., 1972; Novikoff and Novikoff, 1973).

The appearance of the tubule alters gradually during its course through the outer medulla. Soon after its site of origin the epithelium becomes thinner and the interdigitated cell processes reach higher into the cell and extend to the tubular lumen. This results in an elongation of the zonula occludens as shown by the high number of tight junctions met in a longitudinal section. The luminal surface flattens and may possess a few short leaflets or stubby microvilli. The nucleus shifts to the apical half of the cell and often shows invaginations as well as extensive chromatin condensations.

Cortical straight part of the distal tubule (Fig. 31). In the cortical straight part of the distal tubule the epithelial thickness decreases even more strongly than in the medullary part and reaches its minimum near the macula densa. This thin epithelium has generally lost the characteristic features of the medullary part. The lateral cell borders, comprising a narrow intercellular space, are tortuous. The interdigitating cell processes now irregular in shape, size, and arrangement may be parallel to the basal lamina and possess fewer basal villi. The (absolute and relative) amount of mitochondria has decreased. The mitochondrial profiles are smaller and irregular in shape and widely lack the narrow association to lateral cell membranes. The apical cell membrane

Fig. 31, a and b. Distal tubule, straight cortical part; *(a)* cross-section through the end portion of the straight part near the macula densa. The flattened nuclei occupy the total height of the thin epithelium, the mitochondria are irregularly arranged. x ~5500. *(b)* Longitudinal section showing the irregularly shaped interdigitating cell processes and the randomly distributed mitochondria which often lack association with infolded cell walls. Rough endoplasmatic reticulum is abundant often in form of the PTS *(arrow heads);* arrows = junctional complexes; L = lysosome; x ~21000. Insert: tight junction in the straight cortical part of the distal tubule; x ~85000

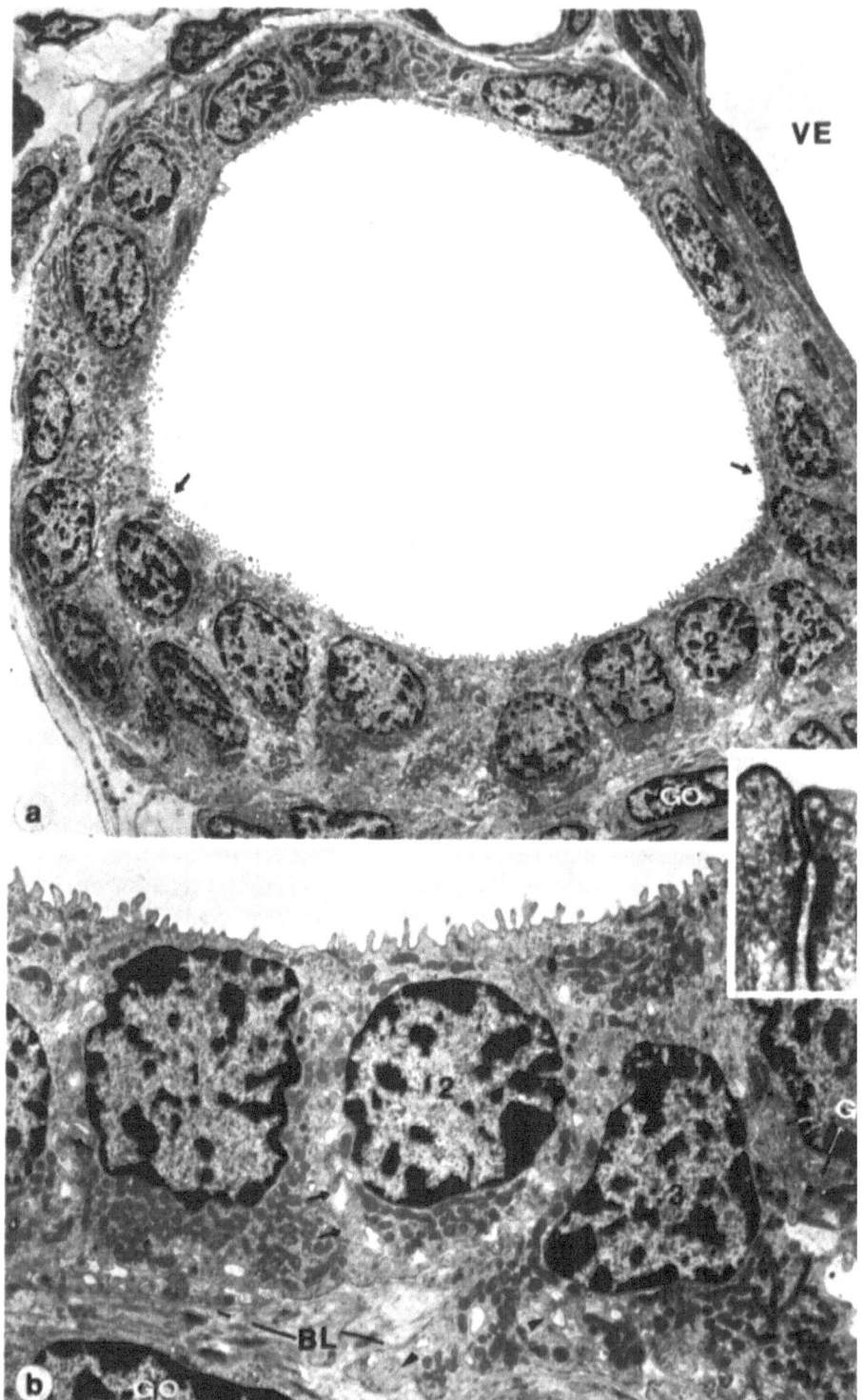

Fig. 32

often has leaflets along the borders of the cells and may possess stubby microvilli in varying numbers. The nucleus has become disclike; it often shows partial segmentations and very dense chromatin condensations. Above and below the nucleus are only small rims of cytoplasm. In the apical rim some vesicles may occasionally occur. The Golgi apparatus is inconspicuous and lies lateral to the nucleus. Long cisternae of the RER in the form of the paramembraneous tubular system and free polysomes are abundant. Lysosomal cell organelles and lipid droplets are rarely present.

Macula densa (Figs. 32, 33). The macula densa in the rabbit is clearly located within the cortical straight part. It forms a plaque of specialized cells at that site of the cortical straight part that touches the vascular pole of the glomerulus. At the opposite side of the macula typical cells of the cortical straight part continue into the short post-macula segment of the cortical straight part. The macula densa cells (Fig. 32) are three to five times as high as the surrounding cells of the cortical straight part. In contrast to the latter, they interdigitate neither with each other nor with neighboring cells of the cortical straight part. The lateral cell membranes possess slender villi projecting into the moderately expanded intercellular space, where they are frequently interlocked in a fingerlike manner. The apical cell membrane is densely studded with slender microvilli. The basis of the cells is not uniformly apposed on a basal lamina, but many irregularly shaped processes protude into an enlarged intercellular space and partially into the basement membranelike material surrounding the Goormaghtigh cells. The macula cells are connected by tight junctions, which are deeper (129.3 ± 23.8 nm, n = 6; see 2.2) than those in the straight part. The cells are stuffed with conspicuously small mitochondria without any association to lateral basal membranes. They accumulate in the basal half of the cell and in the narrow space between the lateral cell membrane and the nucleus. The big nucleus occupies at least the apical two-thirds of the cell. In the thin apical cytoplasmic rim few vesicles can be found. The Golgi apparatus is inconspicuous and situated, as in all other parts of the distal tubule, basal or lateral to the nucleus. Polysomes are scattered throughout the cell, whereas cisternae of RER and other cell organelles as lysosomes, etc., are rarely visible.

Fig. 32, a and b. Macula densa. *a.* Section through the macula densa in the cortical straight part of a juxtamedullary nephron. The extension of the macula is indicated by *arrows.* The macula cells carry numerous short microvilli. Their mitochondria are smaller than those of the neighboring cells. The nuclei are situated apically; *VE* = efferent arteriole; *GO* = Goormaghtigh cell; x ~3000. *b.* Three macula cells indicated in *(a)* at a higher magnification. Cell *1* is cut exactly at the apical-basal axis. The apical position of the nucleus and the basal accumulation of small mitochondria is obvious. Basally the cells are split up into numerous small processes, intertwining with processes of the same cell *(arrow heads)*. Laterally the cells lack interdigitations, but are apposed by fingerlike processes *(arrows);* the intercellular spaces are enlarged in contrast to those between all other distal tubular cells. The basal lamina *(BL)* is continuous with that of the Goormaghtigh cells *(GO);* G = Golgi apparatus. x ~6750. Insert: tight junction between macula densa cells. x ~*85000*

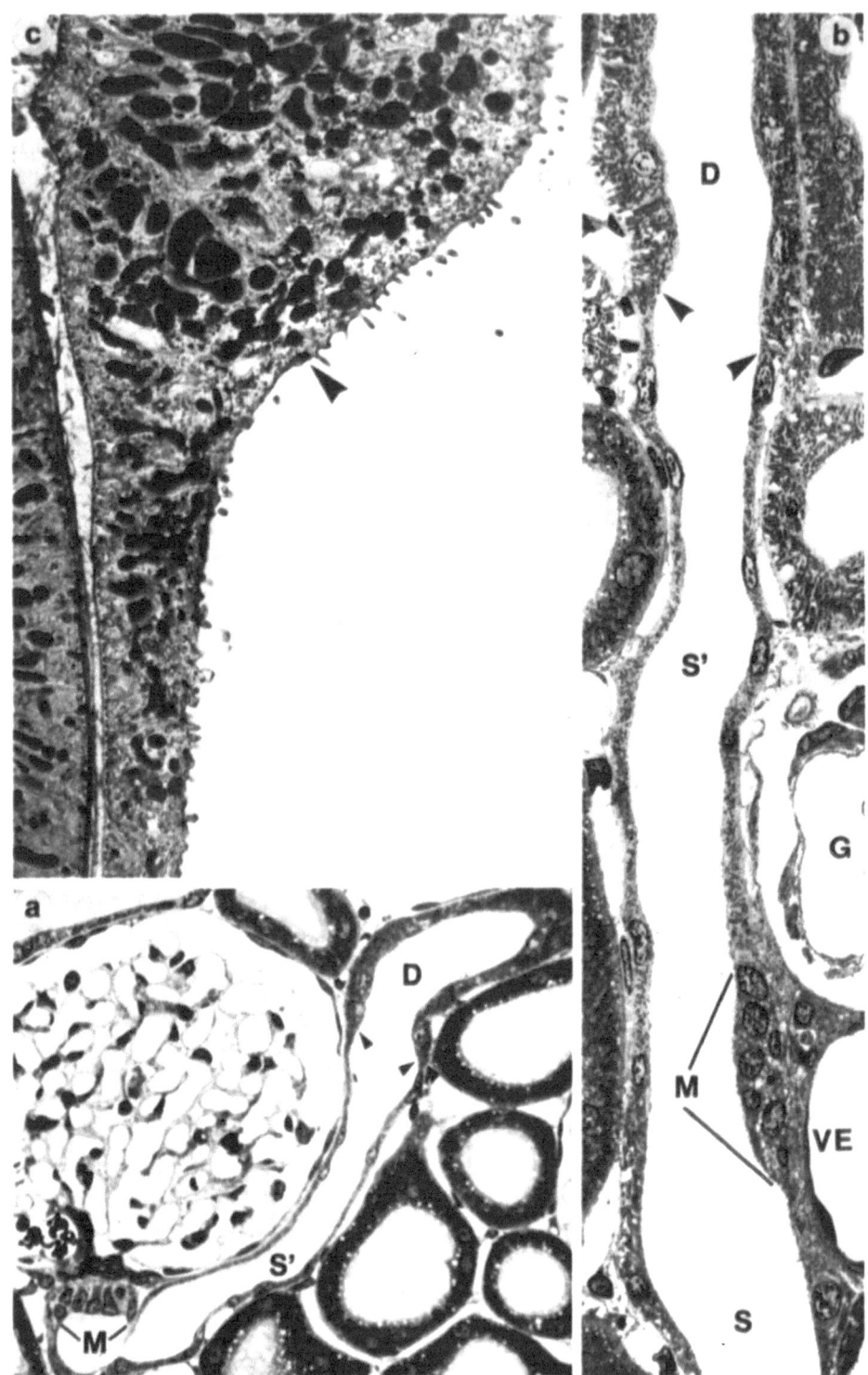

Fig. 33 (Captions to Figs. 33, 34 and 35, see p. 75)

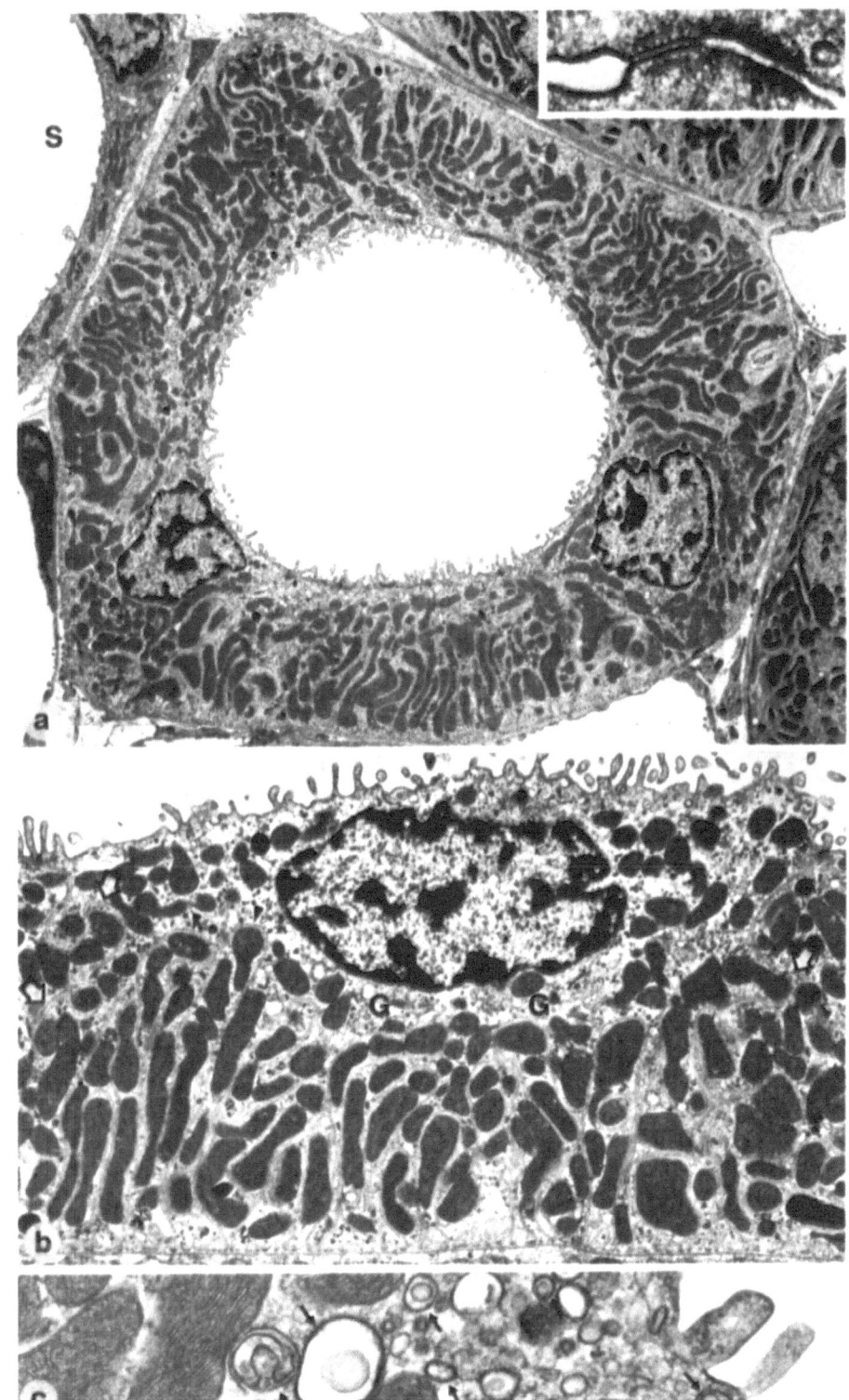

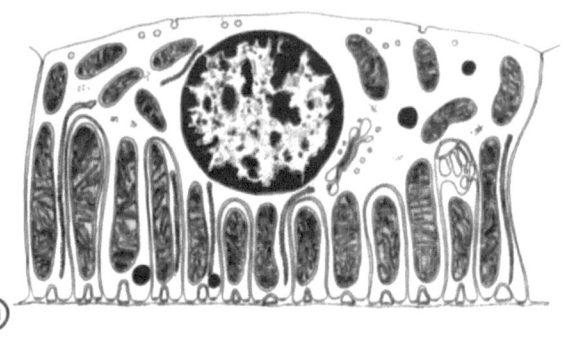

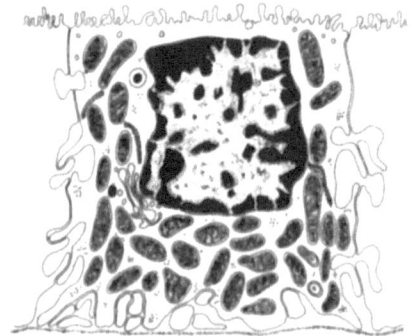

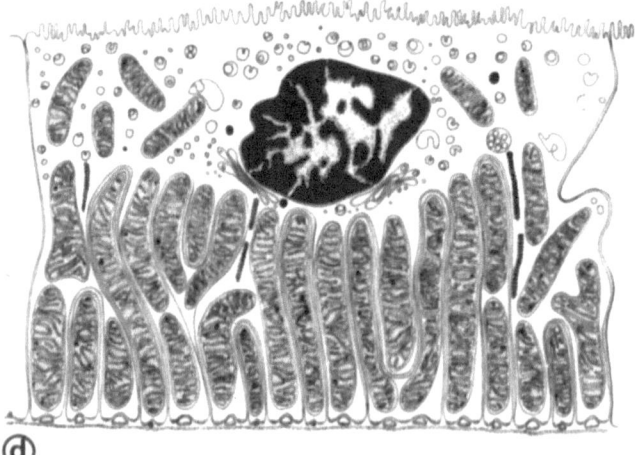

Fig. 35

Convoluted part of the distal tubule (Fig. 33). The beginning of the convoluted part of the distal tubule is sharply demarcated from the foregoing one. No intermediate cell type is interposed between the thin epithelium of the cortical straight part and the three − six times thicker epithelium of the convoluted part. The epithelium of the convoluted part (Fig. 34) is composed of one uniform cell type. The lateral cell membranes are tortuous and fit closely together. Apically the cells are joined by tight junctions, which are distinctly deeper (126.9 ± 24 nm, n = 10; see 2.2) than those in the whole straight part. They are followed by equally deep intermediate junctions. Gap junctions and desmosomes have not been found. The basal three-quarters are extensively interdigitated by cell processes (with small basal ramifications containing filament condensations), which are almost completely filled out with large long mitochondria with a dense matrix and narrowly arranged cristae (intramatrical granules may occasionally occur). Additionally stubby mitochondria are found in the paranuclear region. The apical cell membrane has a dense border of short slender microvilli. The big nucleus, situated in the apical third of the cell, often shows an irregular form and extensive chromatin condensations. The Golgi apparatus lies basal or lateral to the nucleus. Cisternae of RER (as PTS) and polysomes as well as lysosomal cell organelles and lipid droplets are rarely seen between the mitochondria. The upper half of the cell contains numerous vesicles of different sizes, made conspicuous by their invagination of the limiting membrane. They are concentrated under the apical membrane but may also be found in the vicinity of the Golgi apparatus.

Fig. 33, a-c. Transition of the cortical straight part into the convoluted part of the distal tubule. *a.* Semithin section showing the macula densa *(M)*, the cortical straight part extending beyond the macula as so-called post macula segment *(S')*, and the beginning of the convoluted part *(D)*. The transition is indicated by *arrow heads*. x ~ 320. *b.* Low electron microscopic magnification showing the cortical straight part before *(S)* and beyond *(S')*, the macula densa and the beginning of the convoluted part *(D)*. The macula cells *(M)* are interposed in the straight part epithelium only at the side which is in contact with the vascular pole. On the opposite side the epithelium continues into the postmacula segment without any alteration. *VE* = efferent arteriole; *G* = glomerulus. x ~ 750. *c.* Higher magnifications of the abrupt transition *(arrow head)* from the thin cortical straight part epithelium to the thick convoluted part epithelium. x ~6000

Fig. 34, a-c. Convoluted part of the distal tubule; *a.* Cross section. The convoluted part of the epithelium is densely stuffed with mitochondria; its apical cell membrane is studded with short microvilli. The nuclei are located apically. *S* = cortical straight part; x ~3200. *Insert:* tight junction between distal convoluted cells. x ~ 85000. *b.* Higher magnification showing the intense interdigitation by cell processes *(arrow heads)* which occupy three-quarters of the cell height and are completely filled by large mitochondria. The upper quarter is occupied by short mitochondria and the nucleus. The Golgi apparatus *(G)* is clearly situated beneath the nucleus. The lateral cell borders *(thick arrows)* are tortuous. x ~ 7100. *c.* Invaginated vesicles *(arrows)* in the apical cell regions. x ~36150

Fig. 35, a-d. Distal tubular cells: *a.* in the medullary straight part; *b.* in the cortical straight part; *c.* macula densa cells; *d.* convoluted part. x ~4500

4.3.1 Comparative Aspects

Already in 1909 and 1927 Peter described a segmentation of the distal tubule in the rabbit nephron which took structural differences in isolated tubules into account. In the straight part of the distal tubule (which he called thick ascending limb of Henle's loop) Peter distinguished a thick opaque part ("dicker trüber Teil") which toward as well as within the cortex gradually transformed into a light part ("heller Teil"). This transformation from an opaque to a light tubule most readily reflects the decrease in cellular height and density of mitochondria. The opaque part thus corresponds to the medullary straight part and the light part, to the cortical straight part of the distal tubule. At the contact of the tubule with the vascular pole of the glomerulus Peter (1909) observed a widening of the tubular diameter and an accumulation of nuclei, later called "macula densa" by Zimmermann (1933). Peter noted that beyond the macula, the tubule continues with a segment of varying length which has the same light appearance as the straight part before. He called it the "intermediate piece" ("Zwischenstück"). Since this segment does not differ ultrastructurally from the cortical straight part, they are considered to belong to one another. We refer to the latter as the postmacula segment of the cortical straight part. According to Peter (1909, 1927) the subsequent convoluted part (Peter: "Schaltstück" = "intercalated piece") transforms abruptly from the foregoing part and is very short in the rabbit. Peter's descriptions are in full accord with our ultrastructural findings. Moreover, recent functional differences sustain this segmentation. Morel and co-workers (1976) observed that the short, relatively thin postmacula segment reacted in their experiments like the foregoing cortical straight part. It exhibited a sensitivity to parathyroid hormone (PTH) which abruptly ceased at the transition to the convoluted part (Chabardès et al., 1975 a). The convoluted part itself reacted exclusively to calcitonin (Chabardès et al., 1978) and was unresponsive not only to PTH but also to isoproterenol. Both PTH and isoproterenol reacted strongly with the subsequent connecting tubule (see 4.4.3), thereby also establishing a clear-cut functional border between the convoluted part of the distal tubule and the connecting tubule. Thus, the subdivision of the distal tubule of the rabbit nephron into a medullary straight part, a cortical straight part including the macula densa and extending some short distance beyond it, and a convoluted part is well supported by light microscopic, ultrastructural, and functional criteria. This situation may vary in other species with respect to certain characteristics. However, the situation in the rabbit is regarded, at present, to be the best understood basis for comparative considerations.

The gradual alteration of the straight part from a medullary, thick epithelium to a cortical, much thinner epithelium is known also from other species, although it may not be as pronounced. On the light microscopic level it has been described by Sperber (1944) to occur in almost all of the species investigated and in recent investigations on isolated tubules (Rocha and Kokko, 1973; Burg and Stoner, 1974; Imbert et al., 1975 a) it has also been noted. Ultrastructurally the alteration has been confirmed in the rat (Woodhall and Tisher, 1973; Allen and Tisher, 1976; Kriz et al., 1978). The relatively thin postmacula segment has been observed in serveral species: in the dog, pig, sheep, man, and cat (Peter, 1909). Von Möllendorff (1930) also uses the term "Zwischenstück" (intermediate piece) for this segment. Since then it has been clearly detected in isolated tubules (Sperber, 1944; Morel et al., 1976), however, in conventional light microscopy it was generally overlooked. Recently the segment could be recognized in transmission electron microscopy in the rat (Kaissling et al., 1977) and mouse (unpublished results). In scanning electron microscopic investigations in the rat (Allen and Tisher, 1976) it was observed that cells with extensive lateral cell processes resembling those of the ascending thick limbs were still present beyond the macula. The length of this postmacula segment can vary and a correlation of its length with nephron types (deep or superficial ones) does not exist. Generally very short in the rat, the segment can be of considerable length in the rabbit (up to 500 μm).

The convoluted part of the distal tubule is exceedingly short in rabbit kidney (0.5 1.5 mm). This short segment is sharply demarcated at its beginning and in contrast to the other species investigated (man: Tisher et al., 1968; rat: Woodhall and Tisher, 1973; Crayen and Thoenes, 1975; mouse: Clark, 1957; own observation) also sharply delimited at its end. In the rabbit it generally forms one coil, which in superficial nephrons may touch the renal capsule. In deep nephrons the coil is generally somewhat longer and larger. According to Peter (1909) this basic pattern of the convoluted part in the mammalian kidney is also evident in the mouse. In other species (e. g., sheep, dog, cat, and man: Peter, 1909; rat: Sperber 1944) variations of the pattern are due to the greater length of this tubular portion.

In view of structural and functional data, the distal tubule can be subdivided into several well-definable segments. The segmentation is basically equal in superficial, midcortical, and juxtamedullary nephrons. Although less clear than in the rabbit, the subdivision is obvious in other species (rat, mouse). Therefore, we cannot support the opinion that the straight part and the convoluted part of the distal tubule should be viewed as a "single homogeneous segment, briefly interrupted by the special region of the macula densa" (Tisher, 1976).

4.3.2 Functional Aspects

Straight part. The straight part of the distal tubule in the rabbit emerges abruptly out of the thin limbs in the inner stripe and ends just as abruptly. A varying distance beyond the macula densa it is superseded by the convoluted part of the distal tubule located in the cortical labyrinth. This clearly delimited nephron segment exhibits gradual but considerable alterations in ultrastructural organization along its cource through the outer medulla and the cortical medullary ray. It is astonishing that the alterations in ultrastructure are principally the same in juxtamedullary midcortical and superficial nephrons, since one might expect the composition of tubular fluid entering the straight part at the border of inner and outer zones to be different in long- and short-looped nephrons.

At the beginning all straight parts have the same cytologic organization. Ascending through the inner stripe, the straight distal parts of the juxtamedullary nephrons decrease steeply in outer diameter and cellular thickness. Consequently in the outer stripe the thick cross-sectional profiles of superficial and midcortical nephrons may be easily distinguished from the thinner profiles of juxtamedullary nephrons (lying near the vascular bundles). However, a comparison of the straight parts of both nephron types near their ends at the macula densa shows that the tubules of the straight parts of midcortical and superficial nephrons (after ascending through the medullary rays with continuing reduction in diameter and cell thickness) have become even thinner than those in juxtamedullary nephrons. Thus, there are no qualitative differences among the straight distal tubules of different nephron types; consequently, a homogeneous function is to be expected. It is generally accepted that the straight parts of the distal tubule perform a key function in the renal urine-concentrating mechanism. An active reabsorption of salt (driven by an electrogenic chloride pump through a water impermeable epithelium), leaving behind a diluted tubular fluid, has been unequivocally demonstrated in isolated tubule experiments (Burg and Green, 1973; Burg and Stoner, 1974; Kokko, 1974; Rocha and Kokko, 1973). The overall functional characterization of the straight parts of the distal tubule by the term "diluting segment" (Burg, 1976) of the nephron is convincing. As pointed out by Burg et al. (1973; Burg and Bourdeau, 1978), the straight distal tubule is probably the most important site of action of diuretic drugs.

It is obvious that the cytologic machinery for active salt transport (basal-lateral interdigitations and a high amount of mitochondria associated with the interdigitated membranes) is present. The very high Na-K-ATPase activity in the straight part (Schmidt and Dubach, 1969; Schmidt et al., 1975) fits in well with structural and functional findings, although the exact relationship between the Na-K-ATPase and chloride transport is not understood. Compatibility is also evident between the low electric resistance of the straight part (25 Ω cm^2 measured in the cortical straight part, Burg and Bourdeau, 1978; for comparison in the proximal tubule, 7 Ω cm^2 and in the collecting duct, 270 Ω cm^2) and the relatively low apical-basal depth of the tight junctional belt. In freeze-fracture investigations the depth of the tight junctional belt in this segment was found to be somewhat higher (Schiller et al., 1978) than in our measurements. By all means the junctional depth in this segment is higher than in the proximal tubule and in the ascending thin limbs, but is clearly surpassed by all other tubular segments.

The reduction of the salt-reabsorptive machinery toward the cortical straight part (together with an overall reduction in cell height) is parallelled by a reduction in Na-K-ATPase activity (Schmidt and Dubach, 1969). Although the salt-reabsorptive mechanism in both subsegments (medullary and cortical) appears in principle the same, quantitative differences have been uncovered. According to Burg and Bourdeau (1978), the medullary straight part reabsorbs salt more rapidly than the cortical part but cannot create large concentration differences. In the cortical part the transport is slower; however, larger concentration differences can be established between the tubular fluid and the surrounding interstitium. Thus, the bulk salt transport appears to occur in the medullary segment and the cortical part has been supposed to have some reserve reabsorptive capacity (Burg, 1976; Burg and Bourdeau, 1978), which adjusts the final concentration gradient to the surrounding interstitial fluid, thereby guaranteeing a constant low tubular fluid concentration at the macula densa (Kaissling et al., 1977).

In view of their topographic locations, the structural and functional differences between the medullary and the cortical straight part also seem reasonable. Salt reabsorption in the medullary straight part directly generates the outer medullary hypertonicity, whereas salts reabsorbed by the cortical part no longer contribute to the medullary hypertonicity but are drained off into the systemic circulation. This difference does not affect the possibility that both segments (medullary and cortical) contribute to the generation of medullary solute accumulation by elevating the luminal concentrations of urea, as shown in the passive models of the renal concentrating mechanism

(Stephenson, 1972, 1973 a, b; Stephenson et al., 1974; Kokko, 1974; Kokko and Rector, 1972; Jamison, 1974, 1976).

One may assume additional functional differences between the medullary and the cortical straight part of the distal tubule which parallel structural differences. In the medullary straight part of the rat and rabbit a high sensitivity of the adenylate-cyclase-system to vasopressin has been demonstrated (Imbert et al., 1975 b; Chabardès et al., 1978), whereas in the cortical part it has been found to be low. In contrast the sensitivity to parathyroid hormone (PTH) increases along the cortical straight part, whereas in the medullary straight part almost no response to PTH was found (Chabardès et al., 1978). PTH is known to be important in phosphate and Ca-homeostatis; Ca has been shown to be reabsorbed by the straight distal tubule (Rocha et al., 1977). However, it is not known whether there is a linkage between the PTH sensitivity and Ca-transport at this site.

Macula densa. The specialized cell plaque of the macula densa is included in the end portion of the straight part of the distal tubule; a short postmacula segment separates the macula densa from the convoluted part. The macula densa transmits a signal on distal tubular fluid composition to other components of the JGA (Schnermann et al., 1976; Schnermann, 1978). The postmacula segment has been interpreted as protecting the macula densa from effects of some substances secreted into the distal convoluted part (Kaissling et al., 1977).

In the rabbit the macula densa cells are several times higher than the cells of the surrounding straight part. They show a clear, polar organization with the nucleus situated apical and with numerous small mitochondria in the basal cell half. The location of the Golgi apparatus in the basal cell half beneath the nucleus is considered an important distinguishing characteristic of the macula cells from other distal tubule cells (Latta, 1973; Tisher, 1976). However, this does not apply to the rabbit in which the cell polarity of all prismatic distal tubular cells is the same and, consequently, the Golgi apparatus has the same location.

In other respects the organization of the macula cells is quite different from that of ordinary distal tubule cells. Macula cells completely lack lateral interdigitating cell processes. As the collecting duct cells, they are joined by the interlocking of short fingerlike processes that are fully devoid of cell organelles. Consequently in the macula cells the area of the lateral cell membrane surface is smaller than in the other straight part cells. Na-K-ATPase activity has not been found in the macula cells (Beeuwkes, 1975). The mitochondria of the macula cells usually have no association with the basal-lateral cell walls; moreover, their oxidative capacity appears to be much lower than in the surrounding distal tubule cells (Kazimierczak, 1963; Bucher and Kaissling, 1973; Le Hir, 1978). This agrees with the finding that macula cells lack contact to a capillary. Thus, it can be concluded that the reabsorptive capacity of macula densa cells is smaller and different in quality to the reabsorptive capacity in other distal tubule cells (Thoenes, 1961 a).

An important difference might concern water permeability. The whole distal tubule is known to be relatively water impermeable; direct measurements of water permeability of macula densa cells do not exist. Morphologically the macula densa cells behave as collecting duct cells: the intercellular spaces have been found to be expanded to varying degrees, from nearly totally closed to widely dilated. Such a dilatation can never be observed in other distal tubular segments. Thus, in analogy to the collecting duct cells in which an increase in water permeability is combined with a widening of the intercellular spaces (Ganote et al., 1968; Grantham et al., 1969; Woodhall and Tisher, 1973), wide intercellular spaces between macula densa cells may be interpreted to indicate enhanced water flux (Bucher and Kaissling, 1973).

Convoluted part. The convoluted part of the distal tubule, in the rabbit a short but well-demarcated nephron portion, consists of a purely homogeneous cell population. Again no differences emerge from a comparison of superficial, midcortical and juxtamedullary nephrons. Basal-lateral interdigitation (not reaching luminal surface, thus a relatively short, tight junctional belt compared with the straight part) and a very high content of mitochondria narrowly associated to the infolded cell membranes characterize the epithelium. In comparison to the most similarly organized epithelium, for example the medullary straight part of the distal tubule, the reabsorptive machinery (basal-lateral cell membranes, number of associated mitochondria, density of intramitochondrial cristae) is even more extensively developed in the convoluted part. This agrees with the Na-K-ATPase activity, which is the highest of all nephron parts and surpasses the Na-K-ATPase activity of the medullary part still by the factor of 2 (Schmidt and Dubach, 1969, 1971; Schmidt et al., 1975). The association of this enzyme to the basal lateral cell membranes has been demonstrated by an immunoferritin technique (Kyte, 1976). Moreover, an exceedingly high density of oxidative enzymes in these mitochondria may be derived from semiquantitative cytologic studies (Kazimierczak, 1963; Le Hir, 1978).

Physiologic investigations by in vitro perfusion of isolated segments (Gross et al., 1975) have clearly demonstrated sodium chloride reabsorption in the convoluted part, although the rate appears to be only one-quarter of that in the proximal tubule (Giebisch and Windhager, 1973; Hierholzer and Wiederholt, 1976; Kokko et al., 1978). Thus, quantitatively this mechanism does not fully correlate with the extensively developed machinery. However, this comparison might be questionable since sodium chloride reabsorption in the convoluted distal tubule must overcome steep concentration gradients. Also in antidiuresis these gradients in osmolality between inside and outside are maintained, indicating that ADH (antidiuretic hormone) does not influence the water permeability of the convoluted part (Gross et al., 1975). This finding is confirmed by morphologic data (Woodhall and Tisher, 1973) showing that intercellular spaces of the convoluted part do not expand with ADH (as they do in the collecting duct system) as well as by the finding that the sensitivity of the adenylate-cyclase-system to ADH (used as AVP, arginine- vasopressin) is zero in this segment (Imbert et al., 1975 b; Chabardès et al., 1978). Cytochemical studies (Schmidt et al., 1975) have shown that the Na-K-ATPase activity in the convoluted part depends on aldosterone, however, Kokko and co-workers (1978) have not found evidence of influence of mineralocorticoids on the salt reabsorption in this segment. Chabardès and co-workers (1978) have shown that the convoluted distal tubule is a target site for calcitonin, while a linkage to active Ca-reabsorption, which has been shown to occur in the distal convoluted tubule of the rat (Costanzo et al., 1978), has not yet been established.

A further conspicuous feature of the epithelium of the convoluted part is the high amount of invaginated vesicles, which are regularly accumulated in the apical cell half. In the rabbit these vesicles are regularly found in the convoluted part epithelium; in the rat they are already present in the straight part. Although these vesicles may easily be seen in other published micrographs (Griffith et al., 1968; Tisher, 1976) their significance remains unclear. Ito (personal communication) has found the same type of vesicles in the gastric parietal cells. Therefore, one may speculate as to whether these vesicles are involved in hydrogen-ion transport. This speculation is somewhat sub-stantiated by the fact that a strong carboanhydrase activity was histochemically shown in the convoluted part of the distal tubule and in one cell type of the connecting tu-bule (Lönnerholm, 1971, 1973; Rosen, 1972), probably the intercalated cells, where similar vesicles are present. Due to the presence of local glutamic-oxalacetic transamin-ase in the convoluted part it has been suggested that the vesicles could also be in-volved in ammonia excretion (Lee, 1970).

It is especially difficult to make structural-functional correlations in the convoluted part of the distal tubule since data from single nephron preparations are rare (the segment is very short in the rabbit). Physiologic data derived from micropuncture investigations concern species other than the rabbit. In the rat the convoluted part is not a clearly demarcated segment (Osvaldo-Decima, 1973; Crayen and Thoenes, 1975; Tisher, 1976) since the typical convoluted cells are mixed up with those of the subse-quent connecting tubule over a substantial distance. Because all three subportions in the rat touch the real surface, micropuncture data from the socalled distal convolu-tion may well concern different tubular portions, the homogeneous convoluted part, the connecting tubule, the first portion of the collecting duct, or the transitional segment between both. In conclusion, the functional importance of the convoluted part of the distal tubule needs to be more extensively researched.

4.4 The Collecting Duct System

The collecting duct system in the rabbit is described as consisting of the connecting tubule and the collecting ducts (Fig. 5 c). The connecting tubule is a well-definable tubular portion interposed between a distal convoluted tubule and a collecting duct; its beginning and end are clearly delimited by ultrastructural criteria (Figs. 36 and 37). Connecting tubules are situated in the cortical labyrinth. The collecting ducts may be subdivided into the cortical collecting duct (situated mainly in the medullary rays of the cortex and accepting the connecting tubules as tributaries), the outer medullary collecting duct (a totally unbranched part, situated in the outer and inner stripes of the outer medulla), and the inner medullary collecting duct (situated in the inner medulla down to the papillary tip). An inner medullary collecting duct is not a single tube, but is a system of tubes which continuously join together. The term "inner medullary collecting duct" accordingly comprises all the different order seg-ments of this system. Ultrastructurally, the transition between these three collecting duct portions is gradual and cannot be sharply demarcated.

4.4.1 Connecting Tubule

Superficial nephrons drain individually via their own connecting tubule into a cortical collecting duct (Fig. 36 c). The connecting tubule epithelium starts in the descending branch of the tubular coil, which is formed by the convoluted part of the distal tubule (Fig. 36 a, b). In juxtamedullary and most of the midcortical nephrons, which join together to arcades (Fig. 36 b), the beginning of the connecting tubule epithelium is generally located 15 − 20 cells before the joining. Ultrastructurally viewed, the arcades are connecting tubules.

In conventional light microscopy, the transition of the distal tubule to the connecting tubule is not detectable. However, the clear-cut transition between the two is easily recognizable in semithin sections of epon-embedded tissue. The differences in cell structure can be revealed only by electron microscopy.

The connecting tubule (Figs. 38, 39 and 40) consists of two types of cells: the connecting tubule cell and the intercalated cell. The connecting tubule cell only occurs in this tubular segment and thus is typical of it; intercalated cells are also found in the collecting ducts.

In the connecting tubule the ratio of connecting tubule cells to intercalated cells is approximately 5 : 4; moreover, connecting tubule cells surpass intercalated cells in size. Connecting tubule cells can be situated adjacent to each other, whereas intercalated cells are always "intercalated", i.e., separated from each other by connecting tubule cells. Two intercalated cells have never been found to face each other. Cells in the connecting tubule (connecting tubule cells, intercalated cells) do not interdigitate with each other as do proximal or distal tubule cells. Adjacent connecting tubule cells often smoothly appose each other, but may also complexly interlock with short lateral folds. Intercalated cells are apposed to connecting tubule cells by the interlocking of slender fingerlike processes with a frequency of occurrence which increases toward the collecting duct. The tight junctional belt is deep (152 ± 47.6 nm; n = 23), apparently consists of narrowly arranged ridges, and is accompanied by a large zonula adherens. In addition desmosomes occur frequently, even in the very basal part of the epithelium. Gap junctions have not been found.

Fig. 36, a-f (see p. 83). Collecting duct system in the cortex. *Asterisks* = connecting tubules; *A* = arcade; *C* = cortical collecting duct; *D* = convoluted part of the distal tubule; *S'* = post macula segment of the straight part of the distal tubule; *G1* = glomerulus of the superficial layer; *G3* = glomerulus of the third layer; *RS* = renal surface; *V* = interlobular vein; *bars* indicate change of the epithelial lining at both tubule sides; *arrows* indicate urine flow direction; *arrow head* = macula densa. *a.* Section through the superficial cortex showing the transition from the straight part to the convoluted part of a distal tubule and from a distal tubule to the connecting tubule. x ~130. *b.* Section through a midocortical region showing transitions from distal to connecting tubules and fusions of connecting to arcedes. The epithelial lining of the arcades is the same as in the unfused connecting tubules (c-f). Transitions from connecting tubules resp. arcades to cortical collecting ducts. *c.* Cortical collecting duct in the cortex corticis, taking up two connecting tubules. The distinct change in the epithelial lining from the connecting tubule to the cortical collecting duct occurs at the confluence (left), or quite a distance before the confluence (right). *d* and *e.* Cortical collecting ducts in medullary rays taking up an arcade; the epithelium changes a short distance before the confluence. *f.* Cortical collecting duct in a medullary ray taking up an arcade; the epithelial change is at the confluence. In addition, the fusion of two cortical collecting ducts, descending from the cortex corticis is shown. The epithelial lining does not change. *(a) – (f):* 1-μm sections of epon-embedded tissue cut along the longitudinal axis of medullary rays, stained with Azur II and methylene blue. *(a)* and *(b)* x ~130; *(c) – (f)* x ~150

Fig. 37, a-c (see p. 84). Transition of the convoluted part of the distal tubule to the connecting tubule. *a.* Low electron microscopic magnification showing the homogeneous epithelium of the convoluted part of the distal tubule *(D)* and the heterogeneously composed epithelium of the connecting tubule *(Co)*; the transition *(arrows)* is sharp. x ~1100. *b.* In higher magnification the transition *(arrow)* is obvious in the abrupt cessation of the microvilli border, which is present only on the cells of the convoluted part *(D)*. The mitochondria in the connecting tubule cell *(CoC)* are smaller and reveal darker staining than those of the adjacent distal tubular cell. *IC* = intercalated cell; *P2* = proximal tubule *(S2)*. x ~3750. *c.* Part of a distal tubular cell *(D)* adjacent to an intercalated cell *(IC)* of the connecting tubule *(Co)*. The difference between the apical cell membranes is obvious. x ~14200

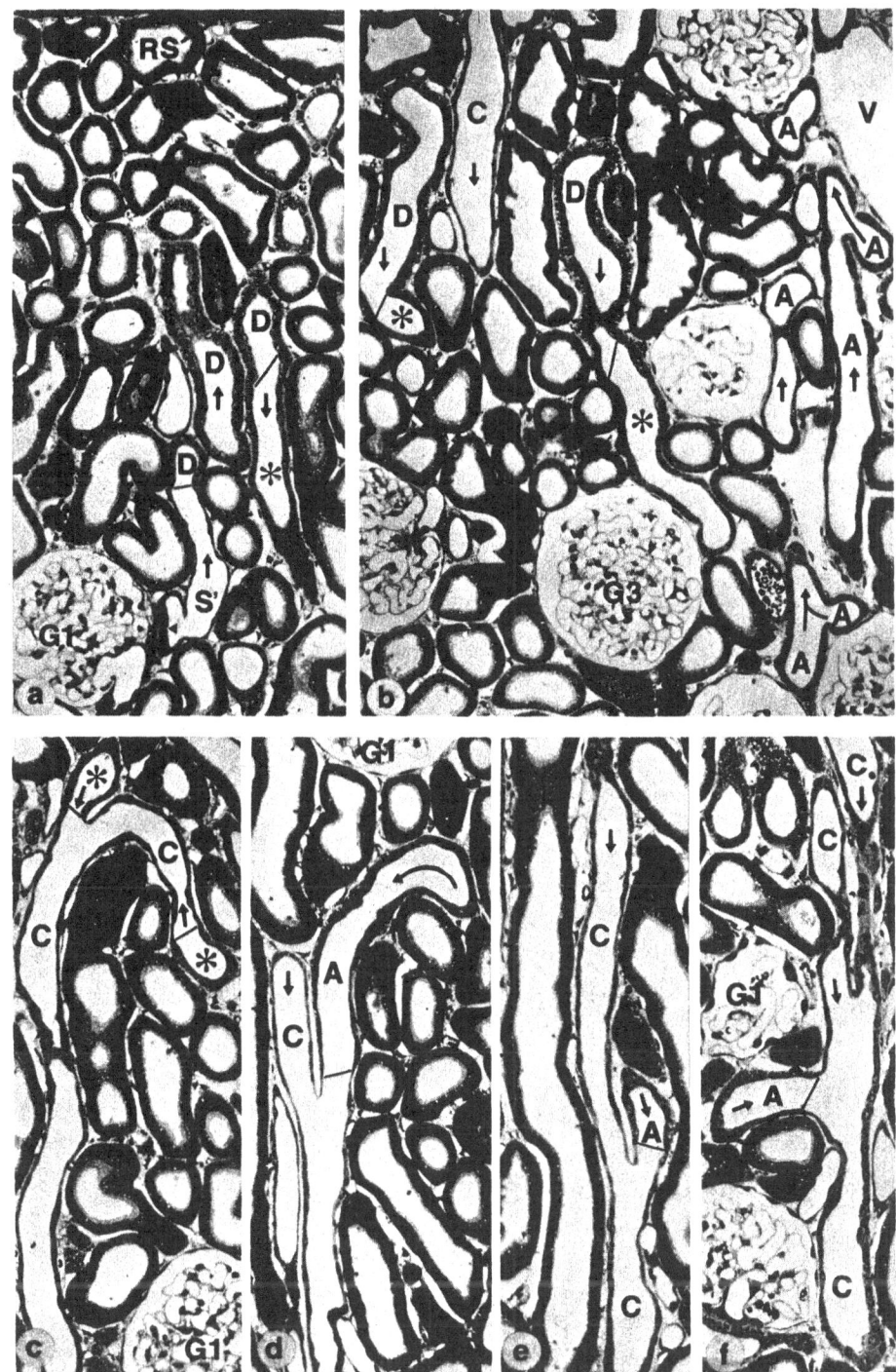

Fig. 36

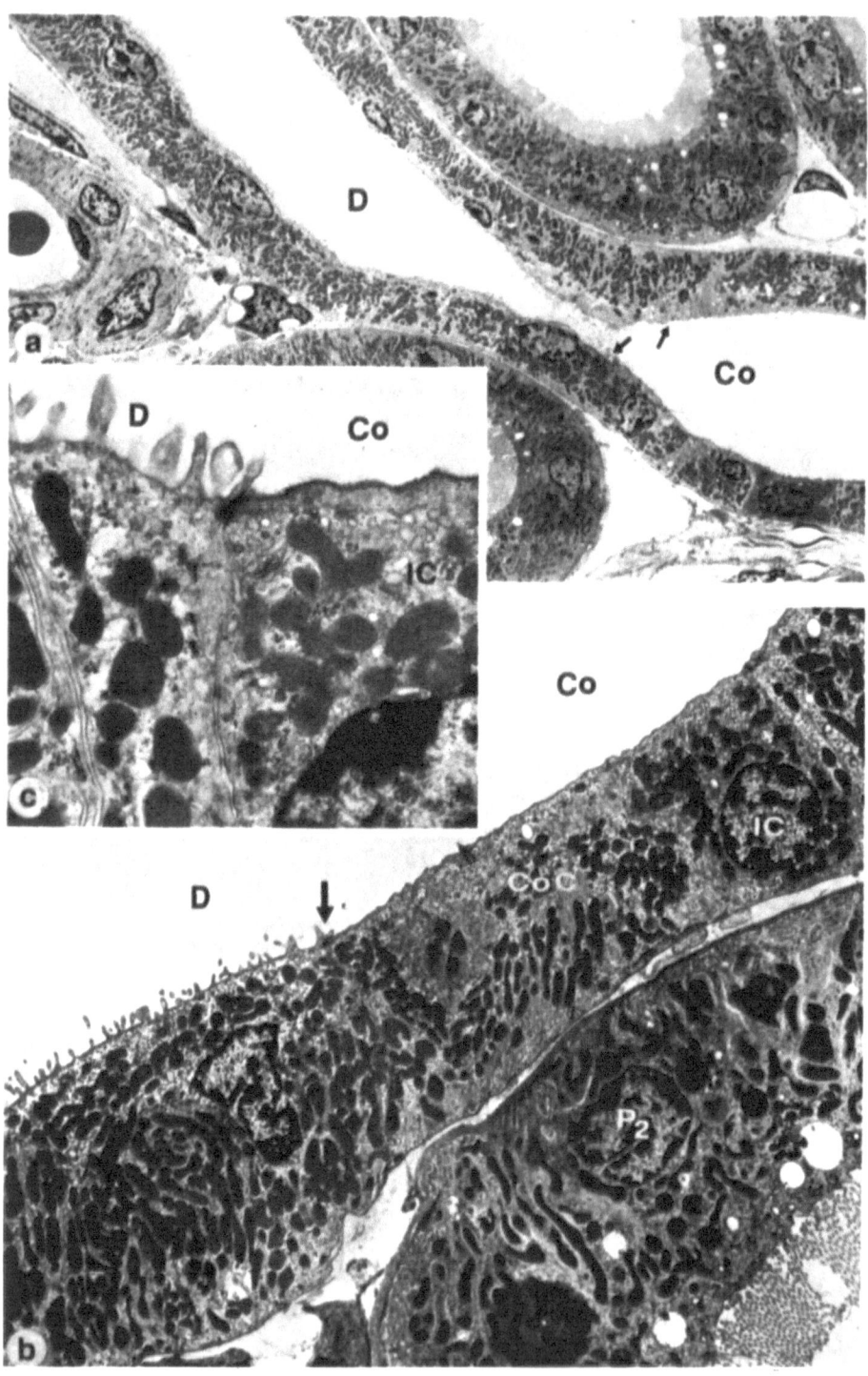

Fig. 37

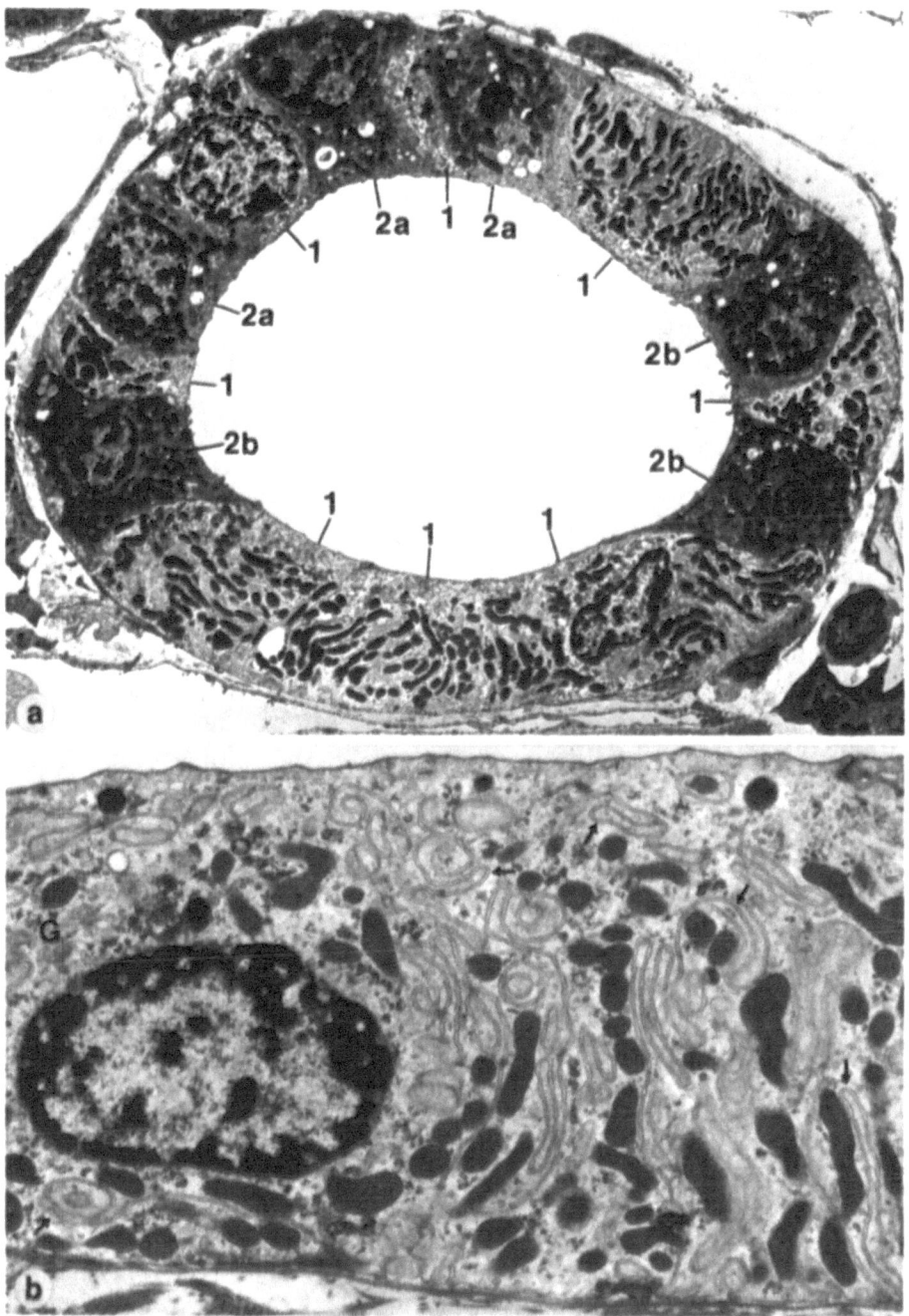

Fig. 38, a and b. Connecting tubule. *a*. Cross section through a connecting tubule showing the different cell types of the epithelium: connecting tubule cells *(1)* and intercalated cells in their "gray" *(2 a)* and "*black*" *(2 b)* manifestation. Intercalated cells are regularly separated from each other by connecting tubule cells. x ~3000. *b*. Connecting tubule cell with light cytoplasmic ground substance and irregularly arranged true infoldings *(arrows)* of the basal cell membrane, which extend into the most apical cell regions. Mitochondria are not systematically associated to the membranes; the apical cell membrane is smooth. *G* = Golgi apparatus. x ~12000

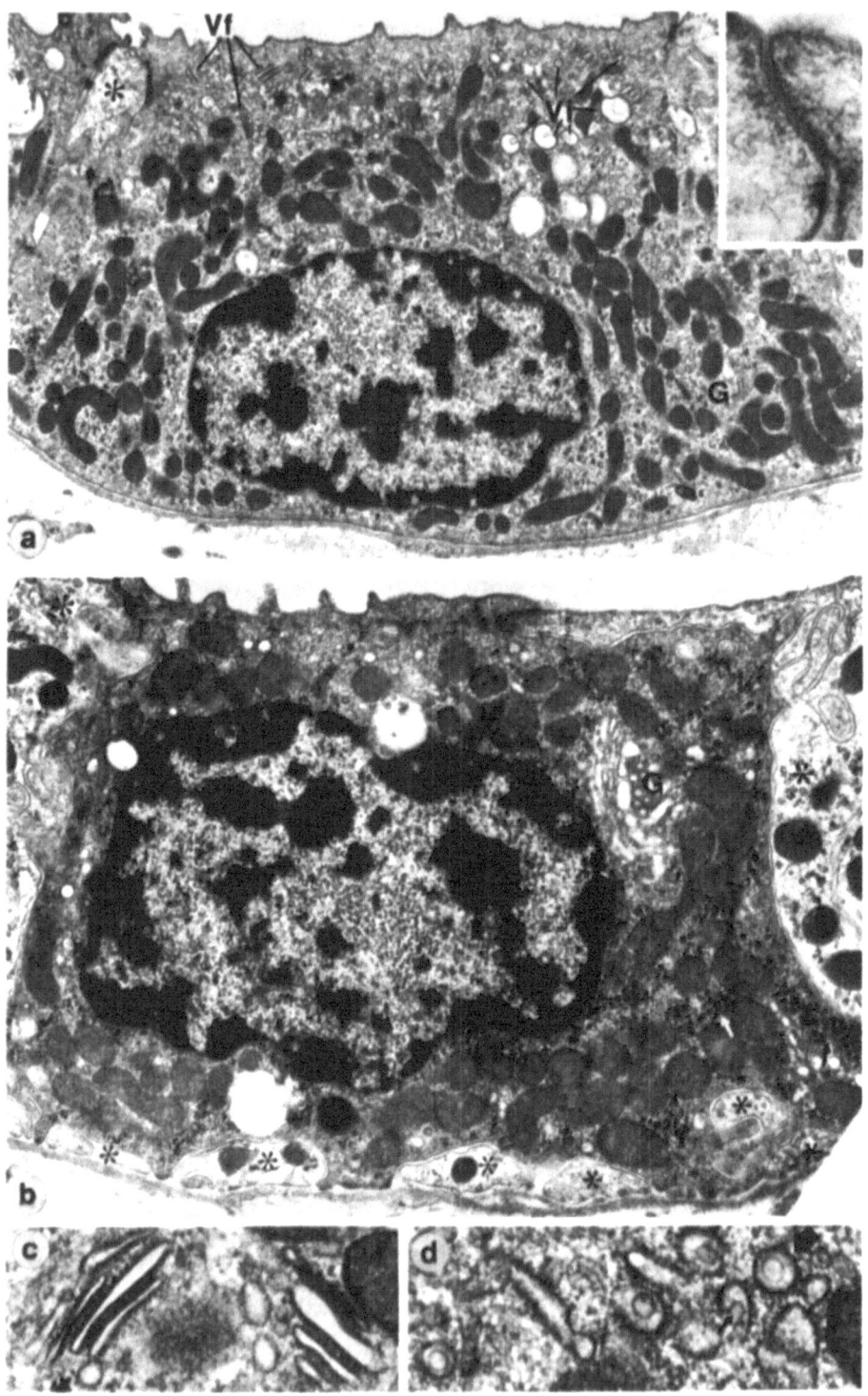

Fig. 39

Connecting tubule cell. In usual electron microscopic specimens the connecting tubule cell (Figs. 39 b and 41 a) seems exceedingly pale. The center of the cell contains a large spherical nucleus. In contrast to the intercalated cells, its luminal membrane is smooth; it may bear a cilia. The most characteristic feature of this cell are the numerous true infoldings of the basal cell membrane (note difference to interdigitations between neighboring cells!) which penetrate into all its regions, including the apical most parts. The infolded membranes are closely apposed and thus the "extracellular" spaces between them are narrow. A regular pattern of these membranes is not obvious and the quantity in which they occur may vary from cell to cell (decreasing toward the collecting duct). Moreover, a regular association with cell organelles is lacking.

The remaining ultrastructural features of this cell are not unique. Mitochondria of a dense matrix (branched forms are frequent) are scattered randomly throughout the cytoplasm. Cisternae of rough endoplasmic reticulum are short and only rarely occur, whereas polysomes are amply distributed all over the cell. The inconspicuous Golgi apparatus is usually situated in the apical cell parts. Lysosomal cell organelles occur but are not prominent.

Intercalated cell. The intercalated cell (Fig. 29) is often termed a "dark cell" due to its overall appearance in light microscopic and usually stained electron microscopic specimens. There are several reasons for the dark appearance of the cytoplasm: high electron density of the cytoplasmic ground substance, high amount of polysomes, and high content of mitochondria with a dense matrix and narrowly arranged cristae intramitochondriales. The degree of darkness does, however, vary. In the connecting tubule of the rabbit a "black" manifestation may be distinguished from a less dark, i.e., a "gray" manifestation, and later a " light" manifestation occurs in the collecting duct.

We therefore prefer the term "intercalated cell." Nevertheless, graduations in the complexly established phenomenon of "cytoplasmic darkness' well characterize manifestations of this cell type. The differences between these variations are never distinct but always gradual with regard to several ultrastructural features.

Accordingly in the connecting tubule the intercalated cell may be described as occurring in two manifestations: a "black" and a "gray;" intermediate forms are also found. Both manifestations are densely stuffed with mitochondria with no obvious association to membranes or other cell organelles. The intrinsic electron density of the cytoplasmic ground substance and the amount of polysomes is less in the "gray" manifestation. The most prominent distinguishing feature between both manifestations are two types of vesicles, both of which are almost exclusively found in the "gray" intercalated cell.

Fig. 39, a-d. Connecting tubule; two manifestations of the intercalated cell. *a.* "Gray" intercalated cell with many mitochondria, a basally located nucleus and a high amount of two vesicle types: invaginated vesicles *(Vi)* and flat vesicles *(Vf)*, accumulated in the apical cell region. The apical cell membrane has an obvious coat; the basal cell membrane is closely apposed to the basal lamina. *G* = Golgi apparatus; *asterisk*-connecting tubule cell. x ~10200. *Inset:* tight junction in the connecting tubule. x ~85000. *b.* "Black" intercalated cell densely stuffed with mitochondria, a high amount of polysomes, especially in lateral and basal cell parts *(arrows)* and a conspicuous Golgi apparatus *(G)*. Vesicles are scarce. The cell is partly separated from the basal lamina by processes of connecting tubule cells *(asterisk)*. x ~17200. *c.* Flat vesicles, typically piled up in groups of four or five, and in perpendicular arrangement. x ~45000. *d.* Invaginated vesicles with longitudinally *(thin arrow)* and transversely *(thick arrow)* cut invaginations. The outer vesicle membrane has a club-shape coat. x ~57000

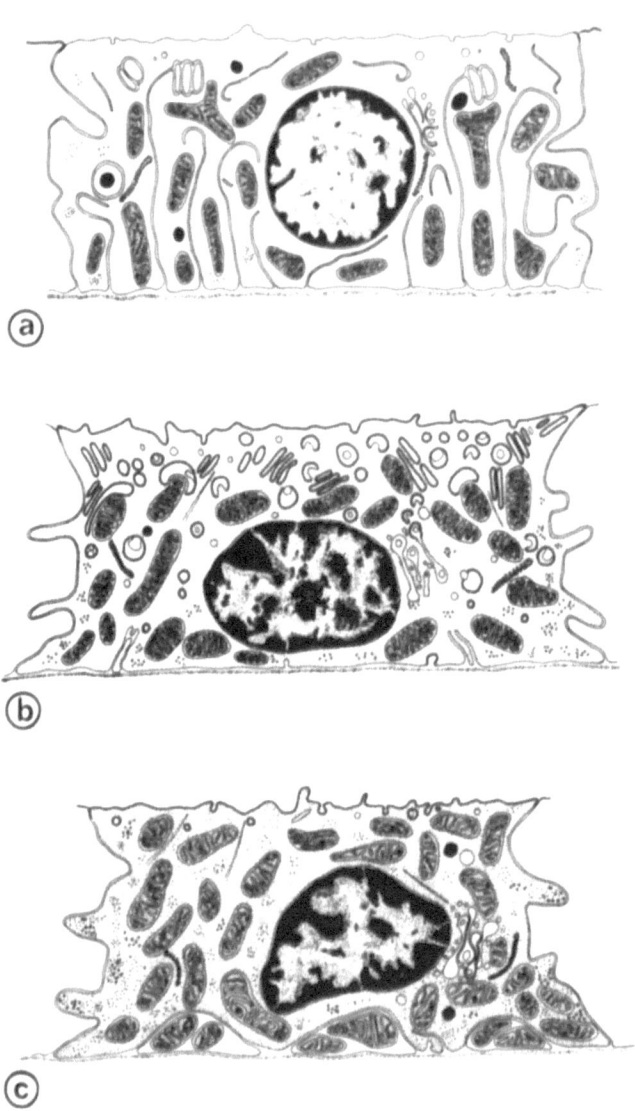

Fig. 40 a-c. Cells in the connecting tubule. *a.* connecting tubule cell; *b.* intercalated cell, "gray" manifestation; *c.* intercalated cell, "black" manifestation. x ~4500

First, spherical vesicles (Fig. 39 d) of different sizes with a coated outer membrane and with bleblike invaginations of a delicate membrane have been found. They are similar but not identical to vesicles encountered in the convoluted part of the distal tubule because the distal tubule vesicles lack the salient coat. Secondly, flat vesicles or saccules (Fig. 39 c) occur which are often piled up to form groups of three-five; the limiting membrane is covered on its outside by a thick coat similar to that of the apical cell membrane. Openings of these vesicles into the lumen have occasionally been found. These vesicular elements dominate the apical part of the cell, while small vesicles of the spherical type may also be found elsewhere in the cytoplasm.

All other features occur in both variations. The luminal membrane bears (increasing in frequency toward the collecting duct) stubby microvilli and is covered by a conspicuous cell coat. A cilia may also be present. The nucleus is basally positioned and has considerable aggregations of heterochromatin. The Golgi apparatus in paranuclear localization is exceedingly well developed with enlarged cisternae, saccules, and many vesicles. It can occupy a large cell area, which is conspicuous by its distinctly brighter aspect. Microtubules are numerous and easily discernable in all parts of the cell. Cisternae of rough endoplasmic reticulum are usually short and occur sparsely.

4.4.2 Collecting Ducts

A collecting duct originates within the cortex corticis near the kidney surface. The site is marked by the ultrastructurally defined transitions of the epithelia where the connecting tubule of a superficial nephron passes over into a collecting duct. Two of those collecting ducts (established only by the transition of one single nephron) usually fuse (Fig. 36 f). This fused duct (eventually also unfused) descending within the medullary ray first (i.e., mostly still in the upper quarter of the cortex) accepts one or two arcades. Subsequently at about the middle level of the cortex, the connecting tubules of those midcortical nephrons not joining an arcade (as do most of them together with the juxtamedullary nephrons) empty individually. In total, one collecting duct accepts an average of six nephrons.

The border between the connecting tubule epithelium (which also outlines the arcades) and the collecting duct epithelium is clearly recognizable (Fig. 36 c–f). As regards superficial nephrons, the border may be situated at the kidney surface and the first part of the collecting duct may touch the surface. In the case of the arcades and the individually draining midcortical nephrons, the collecting duct epithelium generally extends a very short and variable distance (one-four cells) to the arriving tubule (Fig. 36 d and e). The short side branch of the collecting duct often appears thinner in its outer diameter than the arriving connecting tubule. However, the transition from the connecting tubule to the collecting duct can also occur without a side branch and is situated precisely at the entrance (Fig. 36 c and f). The collecting ducts are already fully established at the middle level of the cortex. They descend in a straight manner through the rest of the medullary rays and the outer medulla without fusing. On entering the inner medulla the collecting duct coalesce one by one. At the tip of the papilla only a few large channels open into the renal pelvis.

The collecting duct epithelium reveals considerable, although gradual, ultrastructural alterations along its course from the renal surface to the tip of the papilla. The collecting duct consists of two cell types (Figs. 42, 43 and 44): the principal cell (typical of the collecting ducts) and the intercalated cell (which also occurs in the connecting tubule, see above). In the cortical collecting duct (Fig. 42 a), the ratio between the principal cells and the intercalated cells is about 2 : 1. At the beginning of the outer medullary collecting duct (Fig. 44 a) it is approximately 1 : 1 and increases in favor of the principal cell toward the end of this portion; whereas in the inner medullary connecting duct (Fig. 46 a) only principal cells occur.

The cells of the collecting duct system do not interdigitate by large lateral cell processes, but instead have a simple polygonal-shape outline. The tight junctional belt between the cells (two principal cells or a principal and an intercalated cell may be

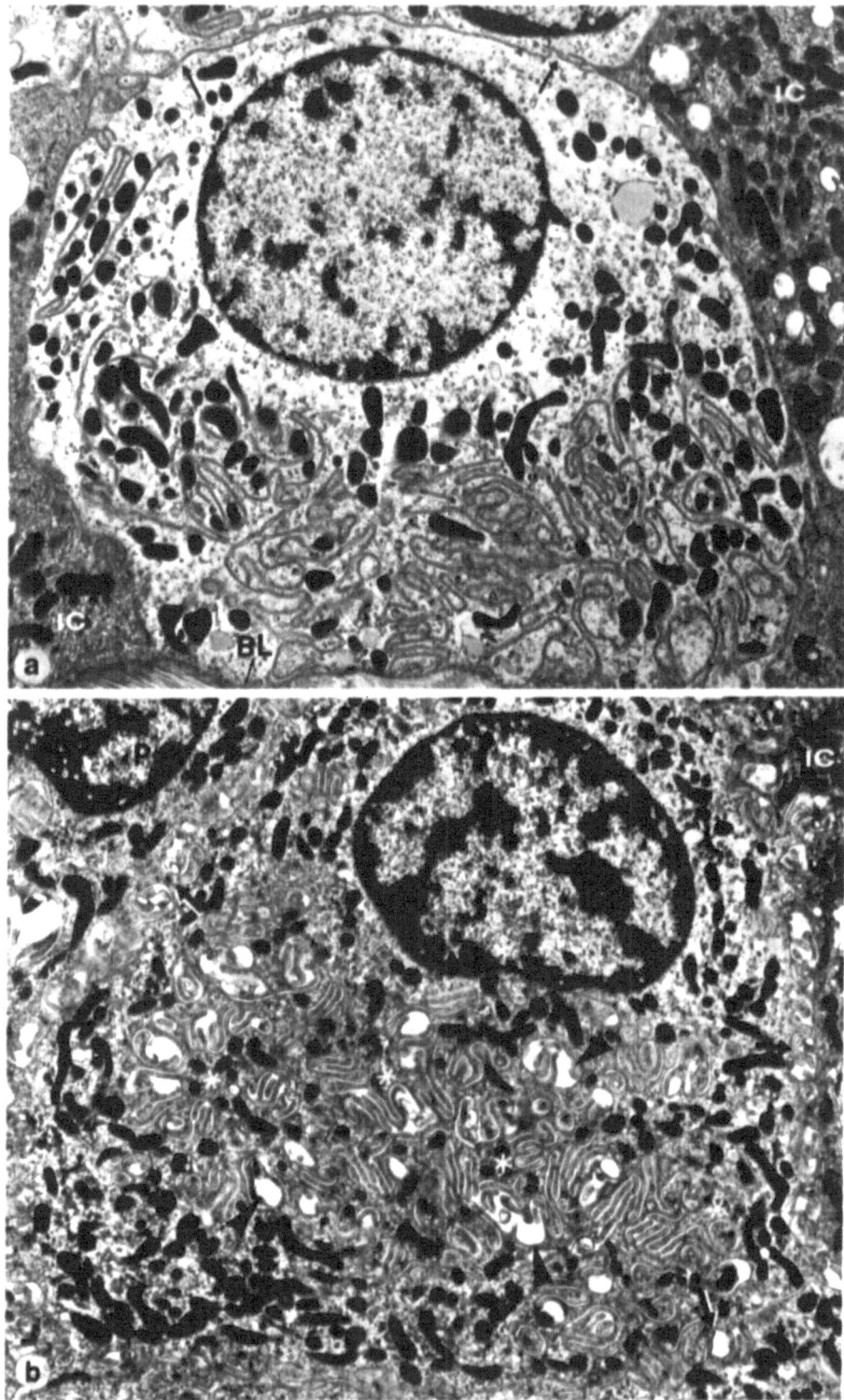

Fig. 41

linked together) is deep (cortical and outer medullary collecting ducts: 191.3 ± 43.9 nm; n = 10; inner medullary collecting ducts: 279.7 ± 73.9 nm; n = 6). The junctional ridges are arranged in a very narrow sequence (in the rat they are impermeable in the cortical and outer medullary collecting duct, yet permeable for lanthanum in the inner medullary collecting duct; Tisher and Yarger, 1975). The tight junctional belt is accompanied by an "intermediate junctional belt" (zonula adherens). In addition, desmosomes are frequently found in the more basal parts of the epithelium; gap junctions are absent.

At the beginning of the collecting ducts (upper half of the cortical collecting ducts) the intercellular spaces between the cells have been found to be slightly dilated (Fig. 42 a). Slender folds and fingerlike cell processes of the lateral walls project into the intercellular space while maintaining a loose interconnection with adjacent cells. In

---

Fig. 41, a and b. Comparison of a connecting tubule cell *(a)* with a principal cell *(b)* of the cortical collecting duct. The differences between both cells are clearly revealed in a section obliquely cut to the apical-basal direction. *a.* Connecting tubule cell, smoothly apposed to another connecting tubule cell *(arrows)* and to intercalated cells *(IC)*. It is evident that the closely apposed membranes within the cells are true infoldings of the basal cell membrane. The cytoplasmic ground substance is pale; *BL* = Basal lamina. x ~8200. *b.* Principal cell with an adjacent intercalated cell *(IC)* and another principal cell *(P)*. Small fingerlike processes *(arrows)* of adjacent cells extend into the slightly enlarged intercellular space and intertwine, but do not penetrate the adjacent cells. The basal part of the cell is split up into numerous stellate-shape parts *(asterisk)* with fingerlike processes intertwining with each other within the same cell. Small cavities of "extracellular" space are often found between them *(arrow heads)*. The cytoplasmic ground substance is denser than that in the connecting tubule cell. x ~7500

Fig. 42, a and b *(see p. 92)*. Cortical collecting duct in the upper cortical half. *a.* Cross section showing the composition of the cortical collecting duct of principal cells *(P)* and intercalated cells *(IC)*. Lateral and basal fingerlike processes extend into enlarged intercellular *(arrow heads)* or "extracellular" *(asterisk)* spaces. *S2* = proximal tubule, segment 2. x ~2800. *b.* Intercalated cell with high cytoplasmic density, a high amount of mitochondria, polyribosomes *(arrows)* and a large Golgi apparatus *(G)*; P = principal cell. The apical cell half contains a high amount of vesicles. x ~10100. Inset *(left)*: In the intercalated cell of the cortical collecting duct small and also large invaginated vesicles are numerous; flat vesicles are less frequent. x ~45000

Fig. 43, a and b *(see p. 93)*. Cortical collecting duct in the lower cortical half. *a.* Cross section showing principal cells with narrow "extra" and intercellular spaces. The stockadelike arrangement of the cell processes in the basal cell half and the distribution of cell organelles in the apical cell half is apparent. One portion of a principal cell *(asterisk)* is obviously darker than other principal cells. One intercalated cell in "black"manifestation *(ICb)* and one in "light" manifestation *(ICl;* the latter can be recognized by the high content of mitochondria and vesicles), lie in the epithelium. *P2-* proximal tubule, segment 2; *S* = cortical straight part of the distal tubule; x ~2800. *b.* Principal cell: the cell organelles are situated in the apical cell half; the basal cell half is split up in numerous slender cell processes which are closely apposed. x ~10500

Fig. 44, a and b *(see p. 94)*. Outer medullary collecting duct. *a.* Cross section in the outer stripe. Principal cells *(P)* and intercalated cells *(IC)* in "light" manifestation are arranged in an alternating pattern. x ~ 2360. *b.* Intercalated cell in "light"manifestation. The high amount of predominantly small invaginated vesicles in the apical cell half is obvious. The number of mitochondria is high and the cytoplasmic density is low compared to that of adjacent principal cells; G = Golgi apparatus; x ~ 10800. *Inset:* apical cell membrane with antennulae microvillares *(arrow head)* and coated invaginated vesicles. x ~45000

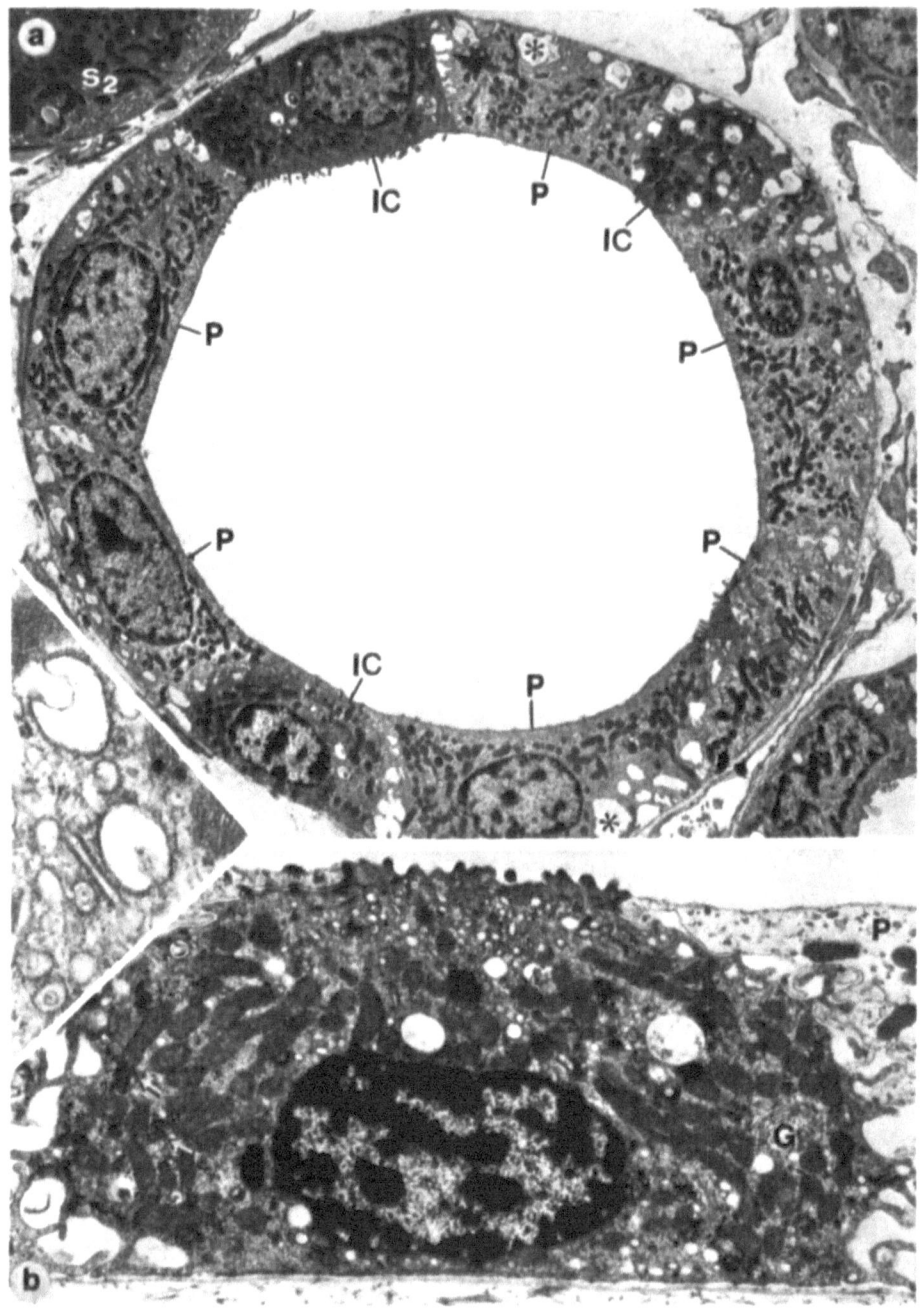

Fig. 42

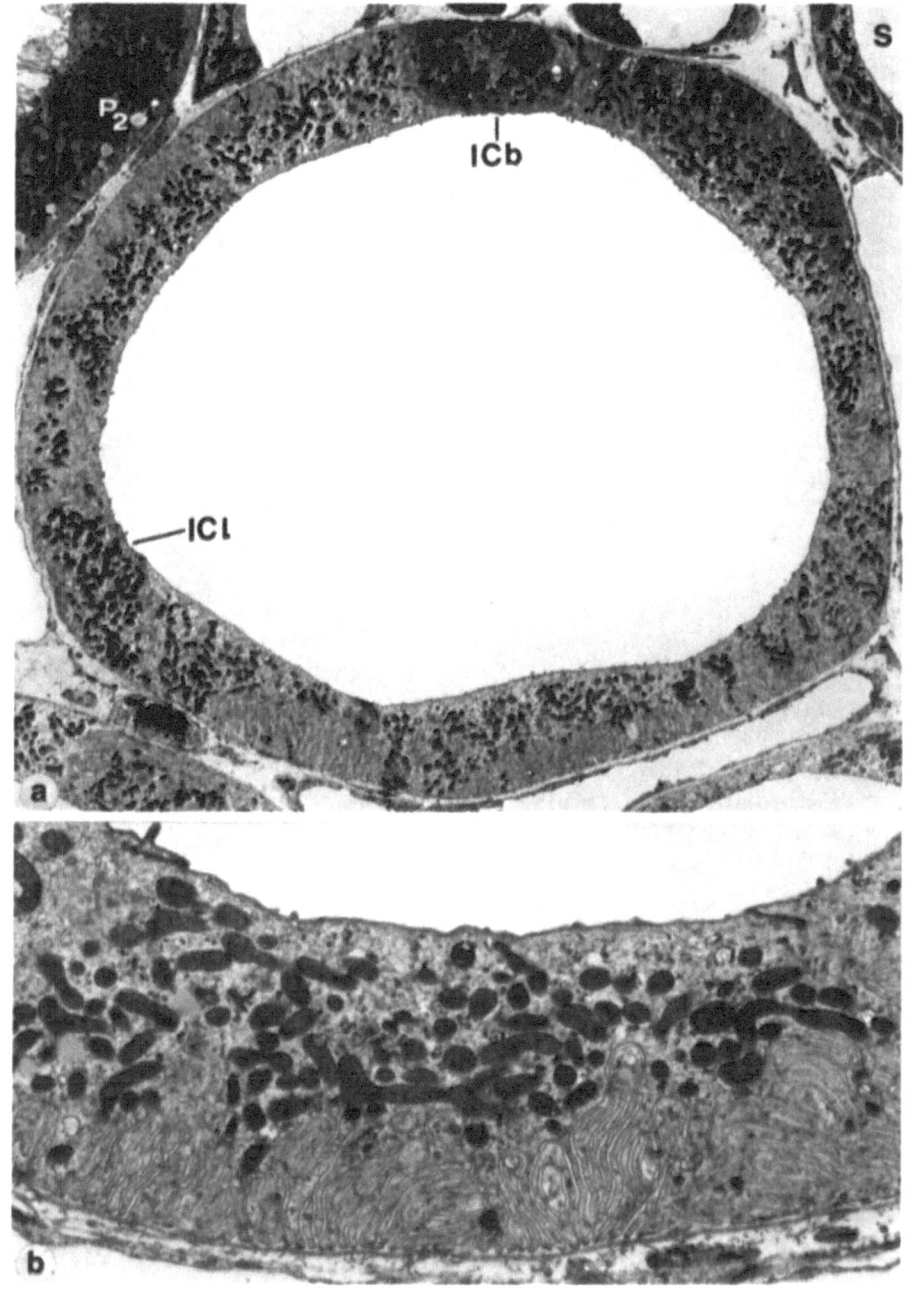

Fig. 43

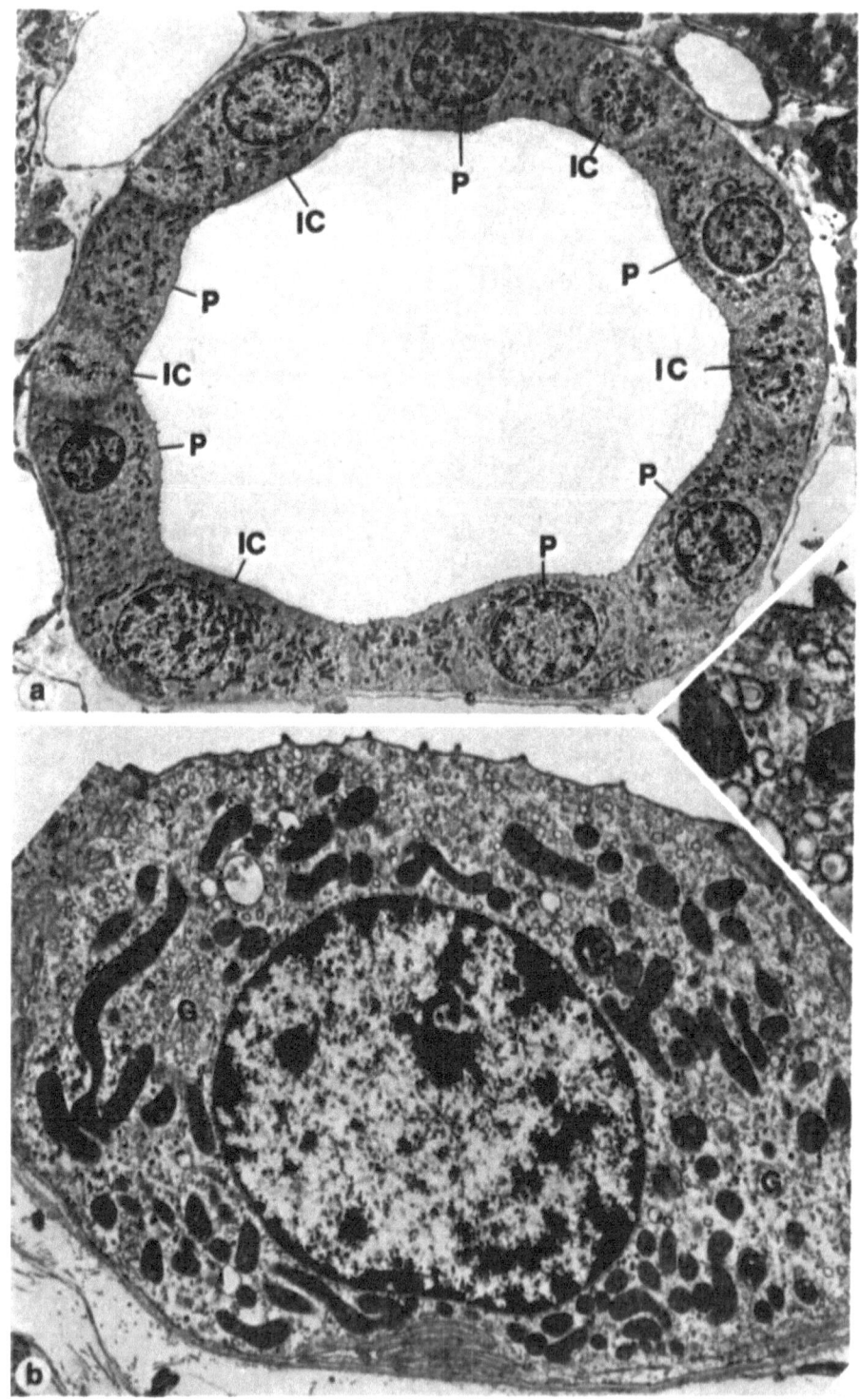

Fig. 44

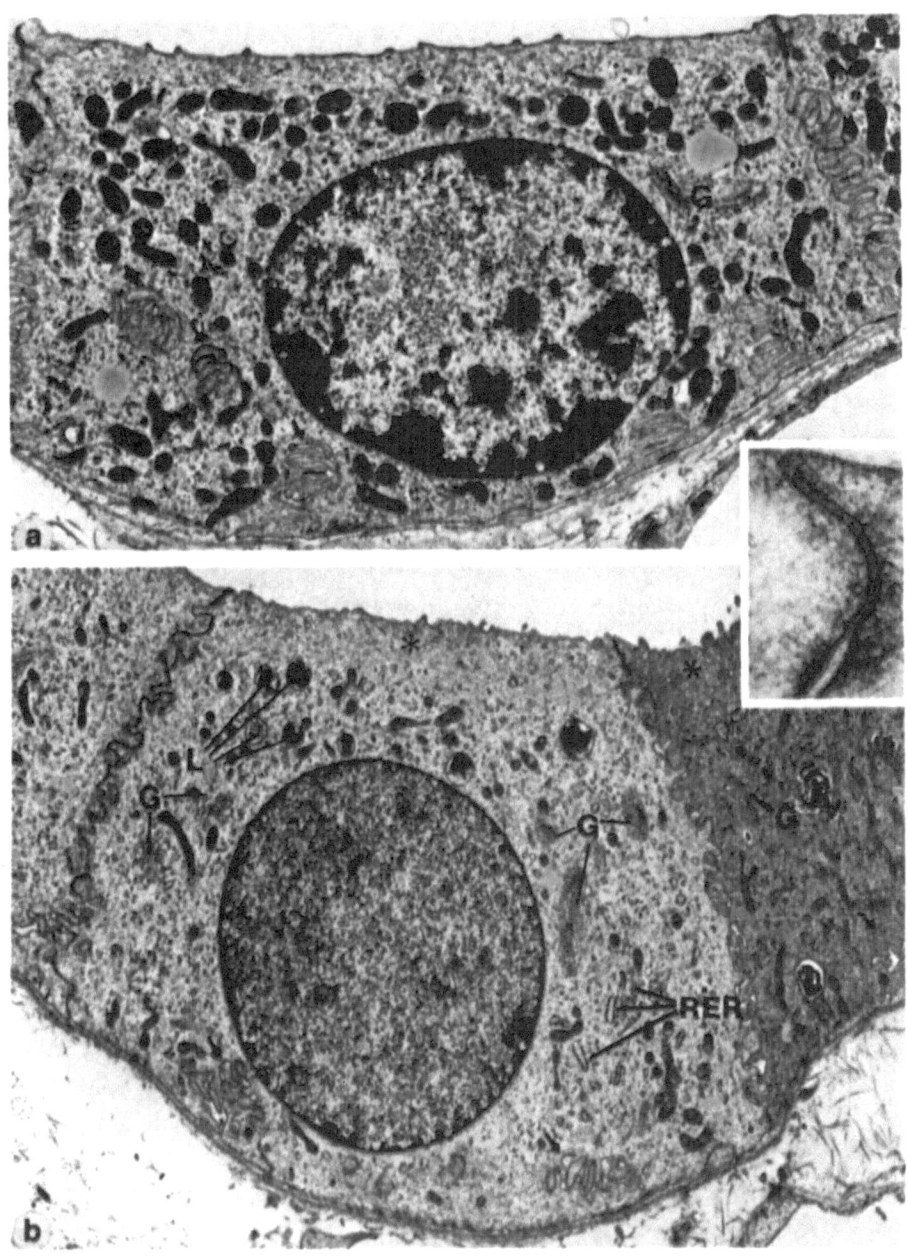

Fig. 45, a and b. Outer medullary collecting duct; principal cells. *a.* Principal cell near the border of outer and inner stripe with randomly distributed cell organelles and a Golgi apparatus *(G)* composed of narrow saccules. The basal labyrinth is limited to only a few areas *(thin arrows)*. Laterally the cells are narrowly interlocked *(thick arrows)*. x ~9100. *b.* Principal cell near the border of inner stripe and inner zone. A few small mitochondria, some lysosomes *(L)*, small, slightly, enlarged cisterns of rough endoplasmic reticulum (RER), and many dictyosomes *(G)* with narrow saccules and small vesicles lie in a light cytoplasmic ground substance. The nucleus shows a homogeneous chromatin structure. The basal labyrinth has reduced to small remnants *(arrows)*. Beneath the apical cell membrane a rim containing a web of microfilaments and microtubules *(asterisk)*, but

y

(contd. p. 97)

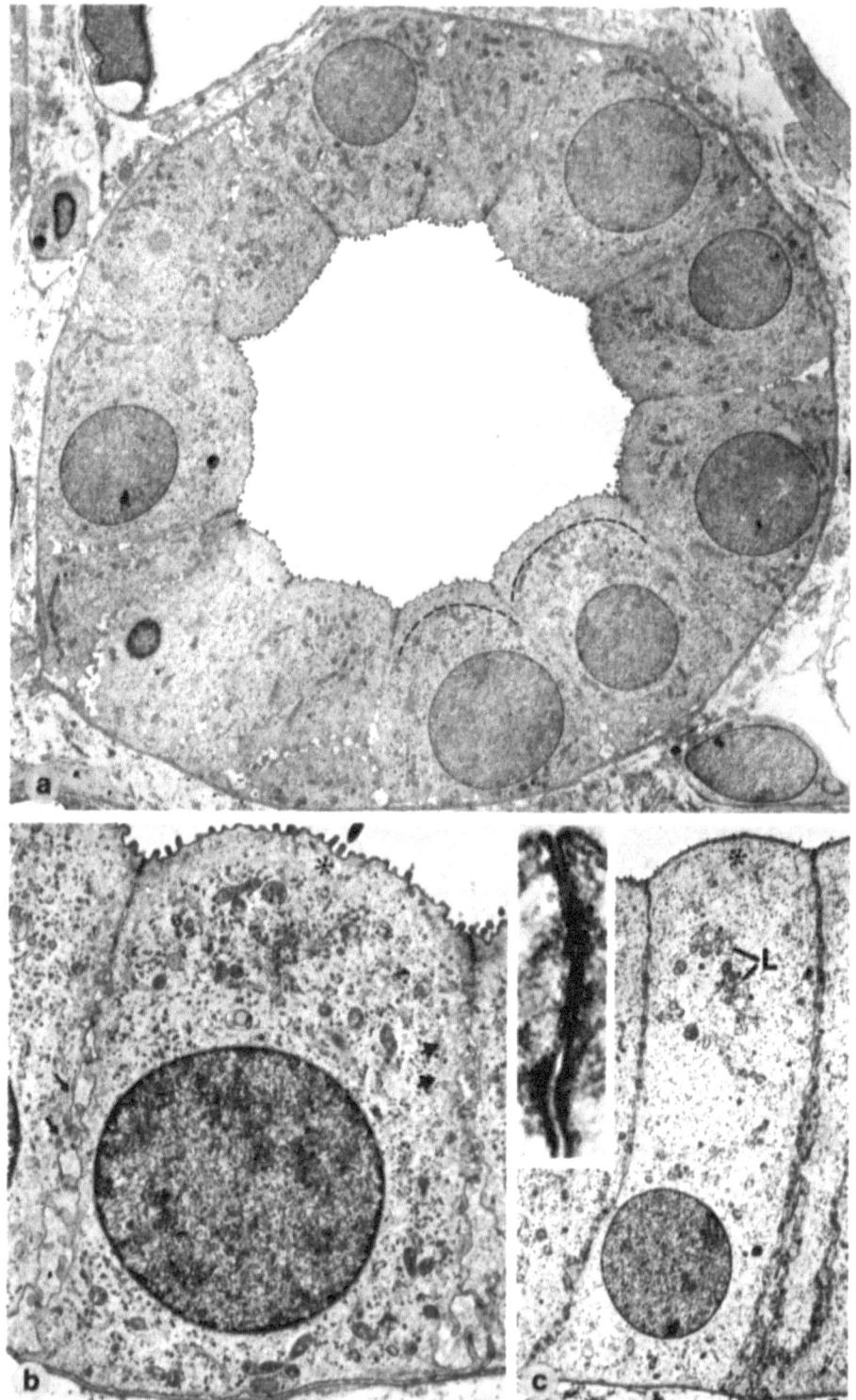

Fig. 46

the lower half of the cortex (Fig. 43 a) and in the outer medulla (Fig. 44 a; outer stripe) the intercellular spaces have been found to be narrow and the cells are densely interlocked with numerous small lateral cell processes. In the inner stripe the intercellular spaces open slightly at the basal parts (Fig. 45 b). At the inner zone the intercellular spaces have been found to be enlarged (Fig. 46 a – c). Stubby lateral cell processes project into the intercellular spaces and are frequently apposed to each other by desmosomes.

Principal cell. The principal cell undergoes gradual alterations from the cortical collecting duct down toward the tip of the papilla (Fig. 41 b; 43 b; 45 a and b; 46 b and c; 47 c–f). Its cell height increases from ~5 – 7 μm in the cortical collecting duct via 10 – 12 μm in the outer medullary collecting duct to 90 – 120 μm at the tip of the papilla. Since the polygonal basis of the cells remains about the same size, the increase in cell volume is at most 20-fold. The spherical nuclei (which in the cortical collecting duct are situated in the upper cell half and in the inner medullary collecting duct occupy a basal position) appear to increase proportionally in volume.

The apical cell membrane has a few blunt microvilli and is covered by a cell coat that is only moderately developed in the cortical and outer medullary portion but conspicuously thick in the inner medulla. Beneath the apical cell membrane microtubules and microfilaments are amply present. They are most conspicuously developed in the outer and inner medullary collecting duct. At these sites they form a dense web which extends like a large rim along the apical and also along the lateral cell membranes in the inner medullary collecting ducts.

Characteristic for the principal cell is the basal "membraneous labyrinth", established by true infoldings of the basal cell membrane. The infolded slender processes are extensively intertwined with each other (all belonging to the same cell!) thus creating a very intricate "extracellular" channel system (Fig. 41b). At the beginning of the cortical cellecting ducts, this "membraneous labyrinth" fully occupies the basal half of the cell, resembling a stockade (Fig. 43 b). Within this labyrinth no major cell organelles are present, the cell organelles are all localized in the apical cell half. Already in the deeper part of the cortical collecting duct, a gradual reduction of this labyrinth becomes evident. In the outer medulla (Fig. 45 a) it no longer occupies the basal part

Fig. 45 contd.
devoid of other cell organelles, is obvious. At the *right*, a dark principal cell is shown. With the exception of its dark, dense cytoplasmic ground substance, its cytologic composition does not differ from that of other principal cells with light aspect. x ~5100. *Inset:* tight junction in the outer medullary collecting duct. x ~85000

Fig. 46, a-c. Inner medullary collecting duct. *a.* Cross section of an inner medullary collecting duct at the beginning of the papilla. The epithelium is composed only of principal cells with large spherical nuclei which exhibit a homogeneous chromatin structure and few cell organelles. The extension of the apical web is indicated in two cells by *broken lines.* x ~ 2400. *b.* Principal cells of the same level as the cross section in *(a).* Note the enlarged intercellular spaces *(arrows)* and the large marginal web *(asterisk)* of microfilaments and microtubules, extending also to the lateral cell walls *(arrow heads);* the basal labyrinth has virtually disappeared. x ~ 3400. *c.* Principal cells near the tip of the papilla. The cells are highly prismatic and poorly equipped with cell organelles. In the apical cell half, lysosomal cell organelles *(L)* in manifold forms have accumulated. x ~ 2400. *Inset:* tight junction in the inner medullary collecting duct; x ~85000

97

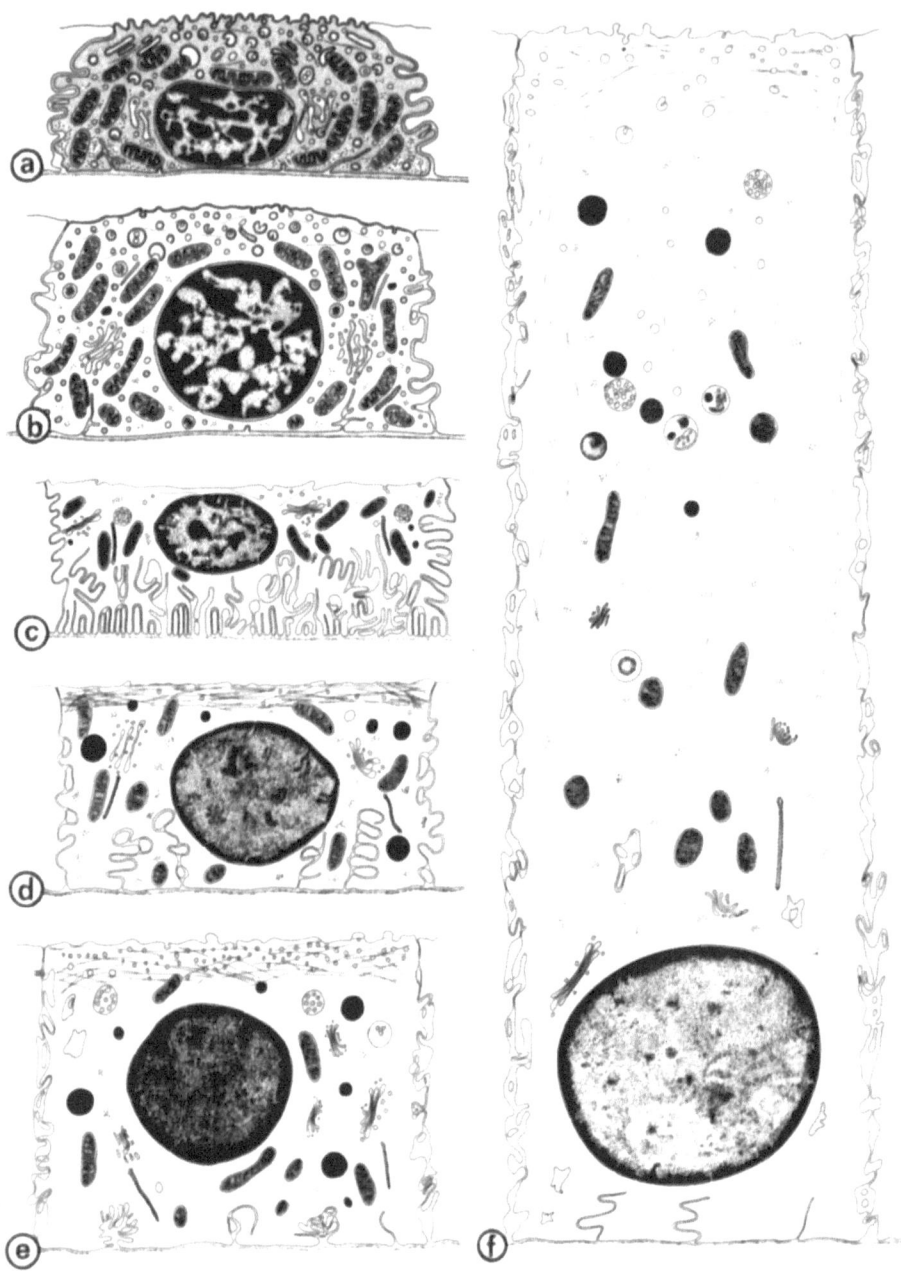

Fig. 47, a-f. Collecting duct cells. *a.* and *b.* Intercalated cells of the cortical (dark manifestation) or outer medullary collecting duct (light manifestation). *c.–f.* Principal cells: *c.* in the cortical collecting duct; *d.* in the outer medullary collecting duct; *e.* in the deep outer and high inner medullary collecting duct; *f.* in the papilla. x ~4500

of the cell but is limited to restricted areas in which shorter folds are more or less narrowly intertwined. In the inner medullary collecting ducts (Fig. 45 b) only small remnants are recognizable at the base of the cell.

Like the intercellular spaces, the "extra"-cellular spaces of the basal "membraneous labyrinth" were dilated in the first portion of cortical collecting ducts. In the deep cortex and in the outer medulla they are narrow, whereas in the inner medulla they are again slightly enlarged.

The cells are amply supplied with short mitochondria (with narrowly arranged cristae and a dense matrix) and polysomes, whereas cisternae of RER are short and sparse. The number of organelles decreases as the distance from the cortex increases. In the cortical collecting ducts the organelles are only found in the apical cell half, whereas in the outer and inner medulla they are randomly scattered throughout the cell.

The Golgi apparatus of the principal cell consists of narrow cisternae and small vesicles. Small Golgi areas, surrounding the nucleus in a beltlike manner, can be encountered in increasing number in principal cells from the outer medulla downward. The lysosomal cell organelles (sparsely present in the cortical collecting duct) also increase considerably in number in the inner zone. In the high prismatic cells of the inner medullary collecting duct agglomerations of multiform lysosomal organelles are generally found in the apical cell half.

In general the overall appearance of principal cells in routinely stained electron microscopic specimens in pale; they have been called "light" cells as compared to dark intercalated cells. However, the cytoplasmic density of principal cells may also vary from cell to cell or even within one cell.

Most frequently in the outer medulla, principal cells with considerably darker cytoplasm can be observed which may by far surpass the cytoplasmic density of adjacent intercalated cells at this site (see below). Features other than the dark cytoplasm have not been found which distinguish them from pale principal cells. However, very dark principal cells are already found in the outer medulla, albeit rarely. They occur more frequently toward the tip of the papilla and obviously represent degenerating stages (showing condensed cytoplasm, nuclear pyknosis).

Intercalated cell. In addition to the occurrence in the connecting tubule (see 4.4.1) the intercalated cell (Figs. 42 b; 44 b; 47 a and b) is only present in the cortical and outer medullary collecting duct. It starts as a very "dark" cell in the cortical collecting ducts and becomes increasingly lighter in appearance within the outer medulla.

Early in the cortical collecting ducts the intercalated cell (Fig. 42 b) corresponds in its cytoplasmic density to its black manifestation in the connecting tubule. Cytologically it approximates the gray manifestation of the connecting tubule, containing a high amount of invaginated and flat vesicles (both coated), a high amount of mitochondria, a large Golgi apparatus, and a basally located nucleus with conspicuous heterochromatin condensations.

The cytoplasmic density of the intercalated cell already diminishes within the deep cortex and transforms into its light manifestations (Fig. 44 b) within the outer medulla. At this site the cytoplasm is less dense than that of the principal cell. The cell size augments and can, in the outer medulla, exceed the size of the principal cell. The flat vesicles diminish and are virtually absent in the inner stripe of the outer medulla. In contrast the invaginated vesicles become plentiful, especially in apical and lateral cell regions, and dominate the aspect of the cell. The mitochondrial content of the intercalated cell remains high. Yet due to the larger cell size, they seem to be less densely packed.

Cisternae of RER are rare and polysomes decrease in number from the cortical to the outer medullary collecting duct. The saccules and cisternae of the Golgi apparatus are prominent at all sites. The content of lysosomal organelles and lipid droplets augments slightly within the outer medulla.

As in the principal cell, microtubules and microfilaments form a dense web beneath the apical cell membrane. This is most prominent in the outer medullary intercalated cells.

The cells are closely apposed to the basal lamina. A few tortuous infoldings of the cell membrane may be present and are encountered in the deeper parts of the outer medulla.

4.4.3 Comparative and Embryologic Aspects

The transition from the nephron to the collecting duct system is not clearly understood. Obviously great species differences exist. In the rabbit a clear subdivision on the basis of ultrastructural criteria is possible: a well-demarcated connecting tubule is interposed between the convoluted part of the distal tubule and the cortical collecting duct. However, the ultrastructurally defined segments do not fully correspond with obvious histologic segments, e.g., an arcade. For precise structural function correlations, the ultrastructural segmentation is most important; nevertheless; the histologic subdivision is justified in certain instances.

The transition from the convoluted part of the distal tubule (homogeneously composed of one cell type) to the connecting tubule (heterogeneously composed of two cell types: the connecting tubule cell and the intercalated cell) is clear-cut. At this point a most characteristic feature of all proximal and distal tubular segments, the interdigitation of the cells by large lateral cell processes, ceases. The cells change from a complicated stellate outline to a simple polygonal outline. The "membraneous labyrinth" of the connecting tubule is predominantly (if not exclusively) established by true invaginations of the basal cell membrane into its own cell; mitochondria are no longer regularly associated with these infolded membranes. Additionally the transition is marked by the abrupt cessation of the short but dense microvilli border of the distal convoluted cells.

The transition from the connecting tubule to the cortical collecting duct is also definite, although the differences are more discrete. Already on the light microscopic level (1-μm thick sections of epon-embedded tissue), the transition is clearly recognizable by (1) a narrowing of the outer tubular diameter and (2) the wide intercellular spaces between the cortical collecting duct cells. In the connecting tubule the intercellular spaces have never been found to be dilated. On the cellular level there are no differences concerning the intercalated cell; this cell type (in various manifestations) is present in both segments, in the connecting tubule and the collecting duct. Yet, the connecting tubule cell and the principal cell of the collecting duct exhibit clear but not striking differences with regard to the "membranous labyrinth" (Fig. 41). In both cells this labyrinth is established by true invaginations of the basal cell membrane. In the principal cell of the cortical collecting duct, the invaginated membranes are regularly arranged like a stockade and only occupy the basal half of the cell; the extracellular spaces between the membranes often appear dilated. In the connecting tubule cell the invaginations do not exhibit a regular pattern and can be found in all parts of the cell even extending into the most apical regions of the cell. The "extra-cellular" spaces between the membranes have always been found to be narrow; thus, the invaginated membranes are always closely apposed to each other.

The transitions from the cortical collecting duct to the outer medullary collecting duct and from the latter to the inner medullary collecting duct are gradual. Therefore the subdivisions are arbitrary. The differences in the segments concern (1) the cell height of the principal cell which increases toward the papillary tip, especially within the inner medullary collecting duct, and (2) the intercalated cells, which have been found to occur most often in the beginning of the outer medullary collecting duct and to be fully absent in the inner medullary collecting duct. Their appearance changes in that within the cortical collecting ducts a dark manifestation predominates, whereas in the outer medullary collecting ducts the "light" manifestation is almost exclusively present. In conclusion, despite differences between the segments of the collecting duct system, common features remain obvious. The cells of all segments have a simple polygonal outline and are joined together by deep tight junctions. Based on ultra-structural criteria, the principal cell is clearly the same cell type in the cortex and in the papillary tip although its cell height in the papillary tip surpasses that in the cortex by up to a factor of 10.

The ultrastructural delimitation of a connecting tubule situated between the distal tubule and the cortical collecting duct corresponds to structural and functional obser-vations on microdissected tubules of the rabbit kidney. In contrast to the foregoing convoluted part of the distal tubule, the connecting tubule reveals a high adenylate-cyclase activity in response to PTH and isoproterenol (Chabardès et al., 1975, a, b). At the transition to the cortical collecting duct, the sensitivity to PTH ceases abruptly, whereas isoproterenol sensitivity continues within the cortical collecting duct (possibly indicating that the intercalated cell, present in both tubular segments, is the target cell of this stimulus; see later). In addition a high sensitivity to AVP was found to start at this transition (see later). Thus, the delimitation of a connecting tubule in the rabbit kidney is well substantiated by morphologic and functional data.

A comparison of findings from the rabbit kidney with those from other species reveals substantial and obvious differences. In the rat (Crayen and Thoenes, 1975), in the mouse (own unpublished data), and in man (Myers et al., 1966; Tisher et al., 1968) a gradual transition from the distal tubule to the collecting duct system has been found. In the transient tubular segment of these species, intermingling of dif-ferent cell types is present. In the beginning, distal tubular cells are mixed with inter-calated cells and in the end, intercalated cells and principal cells form the cortical collecting duct. Whether or not a connecting tubule cell may be clearly delimited, remains unclear. Functional data concerning the adenylate-cyclase activity in the transient tubular segment of the mouse (Chabardès et al., 1978) again support ultra-structural findings: sensitivity to calcitonin (in the rabbit only in the convoluted part of the distal tubule), to PTH and isoproterenol (in the rabbit characteristic of the connecting tubule), and to AVP (in the rabbit only in the collecting duct) has been found.

In summary, the organization of the transition from the nephron to the collecting duct system in the rabbit can in contrast to that in other species, be easily under-stood. Therefore, the organization of the collecting duct system of the rabbit kidney is preferable as a basis for better understanding of the more complicated organization in other species.

The clear ultrastructural findings at this part of the tubular system in the rabbit kidney also add new aspects to the old embryologic question as to where the transition from the metanephrogenic blastema-derived tubule to the ureteral bud-derived tubule occurs in the tubular system. This question has never been unequivocally answered. On the basis of microdissection investigations in the developing human kidney Osathanondh (1963), Potter (1972), and Oliver (1968) contend that the arcades (ultrastructurally the connecting tubule) still originate from the metanephrogenic blastema and localize the embryologic border at or near the fusion of the arcades to the cortical collecting ducts. In his latest publication concerning this problem, Peter (1927) defends the viewpoint that arcades in man and rabbit originate from the ureteral bud. We tend to agree with Peter in localizing the embryologic border in the rabbit at the point where the homogeneously composed distal tubule abruptly passes over into the heterogeneously composed connecting tubule. The following arguments support this viewpoint: 1) The intercalated cell occurs in the connecting tubule and in the collecting ducts, consequently both these tubular portions should be of the same origin; 2) The noninterdigitating connecting tubule cell has many features in common with the principal cell of the collecting duct, but is clearly different from the heavily interdigitated cell of the convoluted part of the distal tubule. Accordingly the two cell types in the connecting tubule are equal or similar to the cells of the collecting duct, whereas they are clearly different from the cells of the foregoing part. Of course these points are evidence and no proof for the development of the connecting tubule from the ureteral bud.

One might ask why we use the term "connecting tubule" instead of the more common term "initial collecting duct." Although we do consider the connecting tubule to be the peripheral most part of the collecting duct system, the term "connecting tubule" has several advantages: 1) it is appropriate for both the branched portion (arcade) as well as for the unbranched portions in superficial nephrons which do not have a "collecting" function; 2) in species with intermingling of cells with possibly different origins in the transient segment from the nephron to the collecting duct, the term "connecting tubule" is far more appropriate; 3) the localization of the connecting tubule within the cortical labyrinth (the arcades take an ascending course within the cortical labyrinth!) suggest a linguistic separation from the collecting duct; 4) the embryologic origin of the connecting tubule from the ureteral bud is not yet proven.

4.4.4 Functional Aspects

The transition from the convoluted part of the distal tubule to the connecting tubule is marked by the following principal changes in cellular organization of the tubules: 1) the homogeneous epithelium of the distal tubule passes over into the heterogeneous epithelium of the connecting tubule; 2) at this point the basal-lateral interdigitation of the epithelial cells with the mitochondria associated to the interdigitated cell membranes ceases. Since the latter ultrastructural feature is generally regarded as the morphologic correlate for substantial salt transport, the overall salt reabsorptive mechanism, or at least its capacity, should also be expected to change at this point too. This statement is corroborated by the significantly lower sodium- potassium ATPase activity in all subsequent tubular segments (Ernst, 1975; Beeuwkes et al., 1975).

However, it has clearly been shown by different techniques and different groups (Diezi et al., 1973; Grantham et al., 1970; Hierholzer and Wiederholt, 1976; Hilger et al., 1958; Jamison et al., 1971; Knox, 1978; Laurence and Marsh, 1971; Sonnenberg, 1976; Stein et al., 1976) that the collecting duct system plays an important role in sodium chloride reabsorption. It is suggested that the final control of sodium excretion by the collecting duct system is moderated by several hormones: aldosterone (Uhlich et al., 1969; Kokko et al., 1978) and ADH (Frindt and Burg, 1974) to stimulate, prostaglandin E 2 (Kokko et al., 1978) to inhibit sodium reabsorption. Moreover, the collecting duct system and the distal tubule are the sites which regulate potassium homeostasis (Giebisch, 1978; Wright, 1976).

In a discussion of structural-functional relationships, the change to a heterogeneously composed epithelium requires differentiation between the cell types. Before discussing the cell types, it should be pointed out that the pattern in which the cells are arranged might well be of functional importance. An intercalated cell (no matter in which manifestation) has never been found to face another intercalated cell. It is always fully surrounded by connecting tubule cells or respective principal cells. It is therefore possible that the cooperation of other adjacent cell types is necessary for some (unknown!) function of the intercalated cell.

It has been suggested (Hagège et al., 1974; Hagège and Richet, 1975; Richet and Hagège, 1975) that the distal tubule cell, the intercalated cell, and the "light" cell of the connecting tubule and collecting duct (i.e., the connecting tubule cell and the principal cell) in fact do not actually represent different cell types, but instead represent different functional stages of one and the same cell. Our research did not support this suggestion as we never found a transitional cell form, which would have indicated the development of an intercalated cell out of the other cell types concerned. On the contrary, we happened to find a mitosis of an intercalated cell ("black" manifestation) in an ultrathin section (unpublished results from the rat kidney), i.e., in that cell type which is supposed by the above mentioned authors always to develop within a functional cycle out of a preceding "light" cell type (distal tubular cell, principal cell). From our experience the distal convoluted cell, the intercalated cell, the connecting tubule cell, and the principal cell are different cell types, although the connecting tubule cell and the principal cell are in many respects similarly organized.

The intercalated cell. The intercalated cell is present in the connecting tubule and in the cortical and outer medullary collecting duct. Intercalated cells have not been found in the inner medulla of the rabbit. However, they may be found in inner medullary collecting ducts of other species (psammomys; own unpublished observation). It has been reported that they increase in number by several stimuli (HCO_3-load, respiratory acidosis: Richet et al., 1970) in the rat and may, e.g., in potassium depletion (Toback et al., 1976), even change their zonal distribution by penetrating into the inner zone.

The intercalated cell occurs in several manifestations, which may well characterize different functional stages. We have already shown that the term "dark cell" for "intercalated cell" should be dismissed since one of its manifestations is conspicuously "light" in appearance. Based on findings in newborn mice by Clark (1957), a "light" intercalated cell might be the source of other cell manifestations, since only the light manifestation is present at birth in mice. The dark manifestations do not occur until several days after birth. Thus, the intercalated cell must be defined by other criteria than its overall appearance, i.e., by the amount and distribution of mitochondria, content of typical vesicles (see above 4.4.1).

We have no functional data to support the view that the different manifestations of the intercalated cell (light, dark, and intermediate forms) are different functional stages belonging to a functional cycle. Objections to the view that dark and light manifestations represent a functional cycle may be based on the distribution pattern of the different manifestations. In the connecting tubule and in the early cortical collecting duct, fully "light" manifestations are absent, whereas in the outer medullary collecting duct they are almost exclusively present. The "light" manifestations may be viewed as an adaptive from to higher interstitional osmolality in the outer medulla, which reacts differently to the fixative than do cells in the cortex. In conclusion, there is a great lack of information about the structural-functional relationships of the different manifestations of intercalated cell.

Even when disregarding different manifestations, no data are unequivocally established about the functional relevance of the intercalated cells. They have been suggested to be involved in HCO_3-reabsorption and H^+-ion secretion (Richet and Hagège, 1975; Richet et al., 1970). The strong carboanhydrase activity in these cells may speak in favor of this function (Rosen, 1972; Lönnerholm, 1971, 1973). The possible correlation of invaginated vesicles to the H^+-ion secretion has already been questioned above (see 4.3.2).

Experiments on isolated rabbit kidney tubules, which measure the adenylate-cyclase activity in response to various hormonal stimuli (Morel et al., 1976; Chabardès et al., 1975 a, b) contribute further to the functional characterization of the intercalated cell. In these investigations the connecting tubule was found to be sensitive to PTH and isoproterenol. When applied together the effects of PTH and isoproterenol became additive. This suggests action on two different sites and possibly on two different cell types. PTH sensitivity ceases at the transition to the cortical collecting duct and sensitivity to isoproterenol continues. Therefore the intercalated cell, which also continues beyond this transition, would be the most probable candidate for the target site for isoproterenol. Because the physiologic relevance of the ß-agonists acting on tubular cells is unknown, conclusions concerning the function of this cell type cannot presently be drawn.

The connecting tubule cell. The connecting tubule cell can be clearly distinguished from the distal tubular cell. A comparison of the connecting tubule cell with the principal cell of the cortical collecting duct shows that both cells have several similar ultrastructural features.

Since in other species (e.g. the rat) the transition from the distal tubule to the collecting duct is gradual, a connecting tubule cell has never been distinguished from a principal cell. However, in the simply organized rabbit kidney they can clearly be distinguished (see 4.1.1; Fig. 41).

Little is known about the specific functional relevance of connecting tubule cells. Experiments concerning adenylate-cyclase-activity in response to hormonal stimuli (Chabardès et al., 1975 a, b) showed that the connecting tubule (in addition to sensitivity to isoproterenol, which is considered to concern the intercalated cell, see above) reacted strongly to PTH with clear-cut borders at the beginning and end of the connecting tubule. Accordingly, the suggestion that the connecting tubule cell possesses the receptors for PTH is justified. This hormone is known to enhance phosphate reabsorption, which corresponds to evidence that phosphate reabsorption occurs in the connecting tubule in arcades as well as in unfused connecting tubules of superficial nephrons (Poujeol et al., 1976, 1977). However, a direct linkage between PTH and phosphate reabsorption has not been established.

The principal cell. The principal cell is the typical cell type of the collecting duct system. Despite tremendous differences in cell height (compare cells in the cortical collecting ducts to those in the papillary tip), the ultrastructural organization remains principally the same. In most other species this increase in cellular height is not as strongly established as in the rabbit. To the best of our knowledge, only the guinea pig's papillary collecting ducts are lined by an equally high epithelium, the functional relevance of which is not known.

Two different kinds of increase in the cytoplasmatic density of the principal cells have been observed. First, in the outer medullary duct, principal cells with a slightly to moderately darker-staining cytoplasm (often only in parts of the cell) are frequently encountered. Their additional features do not differ from a neighboring light-staining principal cell and degenerating signs are not evident. Secondly, in all parts of the collecting duct system with increasing frequency toward the papillary tip single very dark staining principal cells may be found which are clearly identifiable by means of nuclear pyknosis, vacuolization of the cytoplasm, and condensation of mitochondria, etc., as degenerating cells. These two sorts of dark-staining principal cells probably have nothing to do with each other. It is important that they should not be mixed up with "dark cells" in the sense of "intercalated cells". The possibility of such a misinterpretation is given as the intercalated cell occurs in its "light" manifestation in the outermedullary duct. Therefore, principal cells may clearly surpass the intercalated cell at this site in electron density.

The collecting ducts are the site of water reabsorption, a mechanism by which the final concentration of urine is established. This water transport depends on an osmotic gradient between the luminal fluid and the surrounding interstitium as well as on the presence of ADH, which maintains the high water permeability of the collecting ducts (Hays, 1976; Helman et al., 1971; McDonald et al., 1976; Robertson, 1976; Rocha and Kokko, 1974). The principal cell is considered the target site for ADH action. According to Imbert and co-workers (1975 b) the sensitivity to ADH (specifically to AVP since used in their experiments) starts precisely at the transition of the connecting tubule to the cortical collecting duct. The principal cell, which starts at this point, is most probably the only candidate for ADH sensitivity because intercalated cells are already present in the connecting tubule. Since only principal cells are present in the inner zone, the possibility that the ADH action might depend on cooperation of both cell types is ruled out.

The following hypothesis concerning the cellular action of ADH and the transport route of water is now generally accepted. ADH mediated by cyclic AMP increases the water permeability of the luminal cell membrane (Andreoli, 1978). The microfilamentous-microtubular apparatus may be involved in this action, since it has been shown that microtubule-disrupting agents block the ADH effect on water reabsorption (Iyengar et al., 1976; Dousa and Valtin, 1976). Water enters the cell and leaves it via the lateral cell membranes into the intercellular spaces; cellular swelling and widening of the intercellular spaces has been repeatedly demonstrated in response to ADH (Ganote et al., 1968; Grantham and Burg, 1966; Grantham et al., 1969; Tisher et al., 1971; Woodhall and Tisher, 1973; Schafer and Andreoli, 1972). Cellular swelling and especially widening of the intercellular space indicate a bulk water flow; moderate water flow is not expected to cause such large changes in cellular appearance.

Naturally present investigations do not contribute to the understanding of the mechanism of ADH action. However, it has been confirmed that microtubules and microfilaments are well developed in principal cells. They form a dense network beneath the apical cell membrane and become increasingly prominent toward the inner medullary collecting ducts, where this micronetwork even extends to the lateral cell walls. It has to be mentioned that the intercalated cells also contain a substantial number of microtubules and microfilaments.

All our animals at the time of perfusion may be expected to have been in a state of moderate antidiuresis (they had been deprived of food 24 h before killing, but had had free access to water). In all animals a wide opening of the intercellular spaces was found only between the cells of the cortical collecting duct, starting abruptly at the transition from the connecting tubule to the cortical collecting duct. Beginning in deeper parts of the cortical collecting ducts, the intercellular spaces become narrower and are fully closed or only slightly enlarged in the outer and inner medullary collecting ducts. According to the above mentioned findings, a bulk flow of water in our animals may only be expected to occur in the cortical collecting ducts. If a relatively low ADH level in a moderate antidiuresis is assumed, agreement is shown with findings that demonstrate the highest ADH sensitivity to occur at the beginning of the cortical collecting ducts and already begin to decrease along the cortical collecting ducts (Imbert et al., 1975 b; Chabardès et al., 1978). For reasons of efficiency, cortical reabsorption of water (fully utilizing the established gradient in the cortex) is preferable, since it does not lessen the cortical medullary gradient and thus avoids wasting osmotic energy.

4.5 Appendix I: Distal Tubule and Collecting Duct Nomenclature

The different nomenclatures used in the literature are summarized in Table 1. In this paper a nomenclature based on cytologic criteria is applied. Other segmentations are based on functional and histologic criteria. The abbreviations have been introduced in microdissection studies by Morel and co-workers (1976); they are founded on the morphologic aspect under a stereolupe and have the following meanings:

MAL — medullary thick ascending limb;
CAL — cortical ascending limb;
DCTa — distal convoluted tubule, a;
DTCb — distal convoluted tubule, bright;
DTCg — distal convoluted tubule, granular;
DTCl — distal convoluted tubule, light;
CCTg — cortical collecting tubule, granular;
CCTl — cortical collecting tubule, light.

Table 1

SEGMENTATION of the DISTAL TUBULE and the COLLECTING DUCT SYSTEM based on						
CYTOLOGICAL CRITERIA			FUNCTIONAL and HISTOLOGICAL CRITERIA			
DISTAL TUBULE	STRAIGHT PART	MEDULLARY	MAL	THICK ASCENDING LIMB of HENLE'S LOOP		OUTER MEDULLA
		CORTICAL	CAL			MEDULLARY RAY
		MACULA DENSA	DTC a			
		CONVOLUTED PART	DTC b	EARLY	DISTAL CONVOLUTION	CORTICAL LABYRINTH
COLLECTING DUCT SYSTEM	COLLECTING DUCT	CONNECTING TUBULE	DTC g / CCT g	LATE		
				INITIAL COLLECTING TUBULE	ARCADE	
			DTC l	TUBULE	CORTICAL COLLECTING TUBULE	
		CORTICAL	CCT l	MEDULLARY RAY COLLECTING TUBULE		MEDULLARY RAY
		OUTER MEDULLARY		OUTER MEDULLARY COLLECTING TUBULE		OUTER MEDULLA
		INNER MEDULLARY		PAPILLARY DUCT — DUCT of BELLINI		INNER MEDULLA

4.6 Appendix II: Tubular and Epithelial Dimensions

The diameter of tubules and the epithelial thickness have been measured in electron microscopic micrographs. The values obtained are summarized in Table 2.

Table 2. Tubular and epithelial dimensions

Tubular portion	Outer tubular diameter in μm	Epithelial thickness in μm
Proximal tubule		
Segment 1	50–65	12–15
Segment 2	45–55	10–12
Segment 3	<55	<10
Outer medullary portions of Thin descending limbs		
Large profiles	30–40	0.5–0.8
Intermediate profiles	20–30	0.3–0.6
Small profiles (short loops)	15–25	0.2–0.4
Inner medullary portions		
Descending limbs	20–30	0.5–1.5
Ascending limbs	20–30	0.3–1.0
Distal tubule		
Medullary straight part	55–30	12–7
Cortical straight part	30–20	7–1.5
Convoluted part	50–60	15–18
Collecting duct system		
Connecting tubule	35–45	6.5–10
Cortical collecting duct	40–55	5–8
Outer medullary collecting duct	40–55	7–10
Inner medullary collecting duct	55–300	10–120

5 Summary

The following morphologic techniques were used to investigate the rabbit kidney; standard histologic techniques, tracing of single silicone rubber injected nephrons in serial sections, filling of arterial and venous blood vessels with silicone rubber, semithin section techniques of plastic embedded material, and standard electron microscopic techniques.

In many respects, the rabbit kidney is a simply structured kidney with clear divisions between the cortex, outer and inner stripes of the outer medulla, and inner medulla. The renal pelvis fully surrounds the inner medulla and, in addition, with its secondary pouches extensively faces the inner stripe.

The blood vessels of the kidney establish a simple pattern that may be regarded as the basic pattern in mammalian kidneys: there are 28% superficial, 63% midcortical and 9% juxtamedullary renal corpuscles. Juxtamedullary glomeruli give rise to vasa efferentia, which divide into the arterial vasa recta supplying the renal medulla.

Together with part of the venous vasa recta the arterial vasa recta establish the small, but very regularly distributed vascular bundles. All bundles retain their status as primary bundles. The venous vasa recta of the inner medulla join the bundles and traverse the inner stripe within the bundles. The venous drainage of the inner stripe partly is performed (lowermost part) via the bundles and partly (middle and upper part) by venous vessels, which ascend independently from the bundles into the outer stripe. The venous vessels from both sites of origin traverse the outer stripe as wide capillarylike channels. They do not ascend into the cortex, but generally empty at the cortical medullary border into arcuate veins or into the lowermost parts of interlobular veins.

The lymph vessels generally start as lymph capillaries in the periarterial loose connective tissue sheet of interlobular arteries in the cortex and leave the kidney via the arterial route. Lymph capillaries together with the periarterial loose connective tissue establish a common drainage system, which extends along the vasa afferentia to the vascular pole of a glomerulus. This drainage system is important for all substances which act on the intrarenal arteries (renin, kallikrein). The renal medulla has no lymphatic drainage.

The kidney is composed of about 60% long-looped and 40% short-looped nephrons. The short loops of Henle are derived from the superficial glomeruli (28%) and the remainder (12%), from midcortical glomeruli. The bulk of midcortical glomeruli (~50% of all glomeruli) rise to long-looped nephrons. The longest long loops originate from juxtamedullary nephrons. Approximately six nephrons drain into one cortical collecting duct. Thereby the juxtamedullary and deep midcortical nephrons join the collecting duct via arcades, while upper midcortical and superficial nephrons drain individually. In the lower cortex and in the outer medulla fusions of collecting ducts do not occur. They join together in succession upon entering the inner zone and a few very large collecting ducts (papillary ducts) finally open into the renal pelvis.

The nephrons are arranged in a simple pattern. Vascular bundles extend the cortical vascular axis (interlobular artery and vein) into the medulla. The cortical labyrinth surrounds the vascular axis. Parallel and next to the vascular axis, the arcades generally ascend and at the level of the uppermost layer of glomeruli, they loop laterally to empty into a cortical collecting duct within a medullary ray. Each medullary ray contains four collecting ducts and the straight proximal and straight distal tubules of corresponding superficial and midcortical nephrons.

In the medulla the loops of Henle of juxtamedullary nephrons are situated nearest to the vascular bundles and the loops of superficial nephrons most distant from them. Accordingly there are no fundamental changes in their location along the entire route of the loops. The corresponding limbs of a loop of Henle generally run parallel and next to each other. Thin loop limbs never penetrate into vascular bundles. The collecting ducts lie at a distance from the bundles and are intermingled with the loop limbs of superficial and midcortical nephrons. In the inner medulla no constant histotopographic relationships are evident.

The histotopographic relationships of tubules and vessels apparently have the following relevance: possibilities of influence to the arterial input, possibilities of influence to the venous output, possibilities of interaction between different tubular segments and recycling routes.

The ultrastructural organization of the proximal tubule is similar to that of other mammals. It consists of three segments, each of which gradually leads into the other.

S1 is situated in the cortical labyrinth, S2, partly in the labyrinth and the upper parts of the medullary rays, and S3, in the medullary rays and the outer stripe. Accordingly S3 (which is known for its secretory function) is predominantly related to the ascending medullary venous vessels. The transition from S1 over S2 to S3 is characterized by a gradual reduction in cellular interdigitation, in content of mitochondria, and in length of brush order. Peroxisomes are amply present in S2 and somewhat less so in S3. S1 of superficial and midcortical nephrons generally lack peroxisomes, while large peroxisomes occur in S1 of juxtamedullary nephrons. In contrast to other species the tight junctional belt remains shallow throughout the entire proximal tubule.

Because the ultrastructural organization of the thin limbs differs according to nephron types, one must distinguish between thin descending limbs of short loops, thin descending limbs of long loops (subdivided into an outer and inner medullary part), and thin ascending limbs of long loops. As in all other investigated species (e.g., the rat) the diameter of the thin descending limbs of short loops is relatively small. They consist of flat, very simply organized, noninterdigitating epithelial cells and are joined together by junctions of an intermediate apical basal depth (scattered ~120 nm).

The thin descending limbs of long loops are heterogeneous; their diameters in the inner stripe clearly correlate to the actual length of loops. In cross sections through the inner stripe the longest long loops exhibit large cross-sectional profiles; whereas the thin limbs of "short" long loops are intermediate in diameter between the short loops and the longest loops. The epithelium of the descending limbs of long loops is basically equal in all long loops. However, the limbs of the longest loops have a thicker epithelium, a more fully developed labyrinth, and more microvilli in the inner stripe. The cells do not interdigitate and are again joined together by junctions of an intermediate apical-basal depth (scattered ~100 nm). Descending into the inner zone, the cells simplify even more. Only a few microvilli and remnants of a basal labyrinth remain. The tight junctions tend to increase in their apical-basal depth to ~120 nm. In conclusion, the structural organization of the thin descending limbs of long loops is different from that in other species (e.g., the rat, Psammomys). These differences may well account for reported discrepancies in functional data.

A short distance before the bend (or at the bend), the descending epithelium transforms into the epithelium characteristic for the ascending limbs. As in all mammals the epithelium of the ascending limbs is established by flat but heavily interdigitated cells. The processes are connected by junctions of a shallow apical-basal depth (~ 30 nm). Accordingly this epithelium is characterized by extensively developed leaky paracellular pathways.

The distal tubule consists of a straight part beginning at the transition between inner and outer medullary zones and extending a variable distance beyond the macula densa in the cortex (it can be accordingly subdivided into medullary and cortical straight part) and of a convoluted part situated in the cortical labyrinth. A single strongly interdigitated cell type forms the epithelium of the straight part. The cells, which in the beginning are ~ 10 μm high, gradually flatten to ~1.5 μm toward the end near the macula densa. Thereby, the parallel arrangement of the large interdigitating cell processes and mitochondria characteristic of the medullary part becomes irregular; the mitochondria decrease in size and often lack association with lateral cell membranes. In the entire straight part the cells are apically connected by shallow tight junctions (~45 nm). The structural differences between the medullary and cortical

110

straight part are parallelled by functional differences in the salt reabsorptive capacity.

The straight part includes the plaque of the specialized macula densa cells. These cells are four-five times higher than the surrounding straight part cells. They are densely stuffed with very small mitochondria and do not interdigitate with large lateral cell processes. The tight junctions are clearly deeper (~130 nm) than in the surrounding straight part.

The convoluted part of the distal tubule is extremely short in the rabbit and is sharply demarcated from the straight part by a sudden 5 — 6-fold increase in cell height. It consists of one single cell type, which again strongly interdigitates in the basal three quarters of the cells. The high mitochondrial content exceeds that of the straight part. Compared to the straight distal tubule, the zonula occludens is reduced in length by about half and it is increased in depth to ~130 nm.

The collecting duct system is composed of the connecting tubules and the collecting ducts. The connecting tubules are situated in the cortical labyrinth and are interposed as clearly delimited segments between the convoluted part of the distal tubule and a cortical collecting duct. Ultrastructurally viewed, the arcades are connecting tubules; superficial nephrons drain via an individual connecting tubule.

The connecting tubule consists of two cell types: the connecting tubule cell and the intercalated cell, the latter occurs in different manifestations. Both cell types (connecting tubule cell and intercalated cell) have a simple cellular outline and lack intense interdigitations by large lateral cell processes. A tight junctional belt, slightly deeper (~150 nm) than in the convoluted part of the distal tubule, connects them.

The connecting tubule cell is characterized by a membranous labyrinth established by numerous true infoldings of the basal cell membrane. The infolded membranes are closely apposed, reach up into all cell regions, and mitochondria are randomly distributed between the infoldings. All other ultrastructural features are inconspicuous.

A large number of mitochondria, polysomes, flat vesicles, and invaginated vesicles characterize the intercalated cell. Its Golgi apparatus is conspicuously large. In the connecting tubule a "gray" and a "black" manifestation of the intercalated call are present which mainly differ in their vesicular content. While the vesicular content is high in the "gray" form, vesicles are almost absent in the "black" form. Intermediate forms also occur.

The collecting duct is subdivided into the cortical, the outer medullary, and the inner medullary collecting duct. Based on ultrastructural criteria the transitions are gradual. The collecting ducts are lined by two cell types: the principal cell (present in all three collecting duct segments) and the intercalated cell (present in the cortical and outer medullary collecting duct). The cells hava a simple polygonal outline and are connected by a tight junctional belt, which becomes increasingly deeper from the cortical to the inner medullary collecting duct (~190 − 280 nm, respective).

In the cortical collecting duct the basal half of the principal cell is split up into numerous slender folds and processes, which are heavily intertwined. An intricate system of "extra" cellular channels within a cell is thus created. All cell organelles (nucleus, Golgi apparatus, mitochondria) occupy the upper cell half. The aspect of the principal cell changes gradually but considerably. The cell height increases from ~6 μm in the cortical part to between 90 and 120 μm in the papilla. The intricate channel system gradually reduces and displays only a few remnants in the inner zone. The mitochondrial content decreases, whereas that of lysosomes increases. Under the apical cell membrane a web of microtubules and microfilaments becomes increasingly prominent in the inner medulla. In addition to the regular principal cell with rather pale-staining cytoplasm, principal cell with conspicuously dark cytoplasm occur.

In the cortical collecting duct the intercalated cell has a high cytoplasmic density and a large number of invaginated and flat vesicles. Already within the medullary rays, the cytoplasmic density diminishes and the number of flat vesicles decreases. In the outer medulla the cytoplasm of intercalated cells is less dense than that of principal cells. Flat vesicles are virtually absent, whereas the content of invaginated vesicles has increased considerably. In the inner medulla intercalated cells are absent.

The ultrastructural findings suggest that the embryologic border between the nephron and the collecting duct system (derived from the ureteral bud) is situated at the transition from the homogeneous epithelium of the distal tubule to the heterogeneous epithelium of the connecting tubule. This border appears to be of great functional relevance. The separation of the connecting tubule from the cortical collecting duct is also sustained by functional data.

Acknowledgements. The investigations were supported by "Deutsche Forschungsgemeinschaft", grant No. Kr. 546 and by SFB 90, CARVAS, Heidelberg.

The authors wish to thank Ms. I. Ertel for photographical assistance, Mr. A. Laubenthal and Mr. W. Wyrwas for preparing the drawings, Ms. A. Erben and Ms. S. Janeczkowski for secretarial help, and Ms. S. B. Carstens for reviewing the English text.

The quality of the electron microscopic investigations has been decisively determined by the expert technical assistance of Ms. Saliha Šabanović.

Published with the aid of the "Förderungs- und Beihilfefonds Wissenschaft der VG WORT GmbH", München.

6 References

Allen, F., Tisher, C. C.: Morphology of the ascending thick limb of Henle. Kidney Int. 9, 8–22 (1976)

Andreoli, T. E.: A model for the molecular effects of antidiuretic hormone. In: New aspects of renal function (Workshop conferences Hoechst). Vogel, H. G., Ullrich, K. J. (eds.), Vol. VI, pp. 159–174. Amsterdam, Oxford: Excerpta Medica 1978

Ashton, K., Koepsell, H.: Measurement of Na-K-ATPase activity in segments of proximal tubules from superficial and juxtamedullary rat nephrons during antidiuresis. Pflügers Arch. 363, 251–253 (1976)

Aukland, K.: Renal blood flow. In: Kidney and urinary tract physiology II. Thurau, K. (ed.), Vol. 11, pp. 23–79. Baltimore, London, Tokyo: University Park Press 1976

Bankir, L.: Hétérogénéité des néphrons chez le lapin. Etude vasculaire et hémodynamique. Thèse, Université Paris-Sud, Centre d'Orsay 1976 a

Bankir, L.: Vascularisation renale: organisation des capillaires du flocculus. Irrigation post-gloméru-laire. J. Urol. Néphrol. (Paris) 9, 721–766 (1976 b)

Bankir, L., Farman, N.: Hétérogenéité des glomérules chez le lapin. Arch. Anat. Microsc. Morphol. Exp. 62, 281–291 (1973)

Bankir, L., Rouffignac, C. de: Anatomical and functional heterogeneity of nephrons in the rabbit: Microdissection studies and SNGFR measurements. Pflügers Arch. 366, 89–93 (1976)

Bankir, L., Rouffignac, C. de, Grünfeld, J. P., Sabto, J., Funck-Brentano, J. L.: Single glomerular blood flow and single nephron glomerular filtration rate in the hydropenic rabbit kidney. In: Radionuclides in Nephrology. IIIrd International Symposium Berlin, 1974. zum Winkel, K., Blaufox, M. D., Funck-Brentano, J. L., (eds.), pp. 1–8. Stuttgart: Thieme 1975

Bankir, L., Rouffignac, C. de, Kaissling, B., Kriz, W.: The structural organization of the vasculature in the Psammomys kidney. Anat. Embryol. (in press) (1978)

Barajas, L., Latta, H.: A three-dimensional study of the juxtaglomerular apparatus in the rat. Light and electron microscopic observations. Lab. Invest. 12, 257–269 (1963)

Barajas, L., Silverman, A. J., Muller, J.: Ultrastructural localization of acetylcholinesterase in the renal nerves. J. Ultrastruct. Res. 49, 297–311 (1974)

Barajas, L., Wang, P.: Demonstration of acetylcholinesterase in the adrenergic nerves of the renal glomerular arteries. J. Ultrastruct. Res. 53, 244–453 (1975)

Barger, A. C., Herd, J. A.: Renal vascular anatomy and distribution of blood flow. In: Handbook of Physiology. Orloff, J., Berliner, R. W., Geiger, S. (eds.), Vol. VIII, pp. 249–314. Washington: Am. Physiol. Soc. 1973

Bargmann, W., Krisch, B., Leonhardt, H., Malyusz, M.: Lipids in the proximal convoluted tubule of the cat kidney and the reabsorption of cholesterol. Cell Tissue Res. 177, 523–538 (1977)

Barrett, J. M., Majack, R. A.: The ultrastructural organization of long and short nephrons in the kidney of the rodent (Octodon degus). Anat. Rec. 187, 530 (1977)

Barrett, J. M., Kriz, W., Kaissling, B., Rouffignac, C. de: The ultrastructure of the nephrons of the desert rodent (Psammomys obesus) kidney. I. Thin limb of Henle of short-looped nephrons. Am. J. Anat. 151, 487–498 (1978 a)

Barrett, J. M., Kriz, W., Kaissling, B., Rouffignac, C. de: The ultrastructure of the nephrons of the desert rodent (Psammomys obesus) kidney. II. Thin limbs of Henle of long-looped nephrons. Am, J. Anat. 151, 499–514 (1978b)

Beard, M. E., Novikoff, A. B.: Distribution of peroxisomes (microbodies) in the nephron of the rat. J. Cell Biol. 42, 501–518 (1969)

Beeuwkes, R.: Macula densa: Absence of transport ATPase. Kidney Int. 8, 467 (1975)

Beeuwkes, R., Rosen, S.: Renal sodiumpotassium adenosine-triphosphatase: optical localization and x-ray microanalysis. J. Histochem. Cytochem. 23, 828–839 (1975)

Bell, R. D., Keyl, M. J., Parry, W. L.: Experimental study of sites of lymph formation in the canine kidney. Invest. Urol. 8, 356–362 (1970)

Bonvalet, J. P., Bencsath, P., Rouffignac, C. de: Glomerular filtration rate of superficial and deep nephrons during aortic constriction. J. Physiol. 222, 599–605 (1972)

Bourdeau, J. E., Carone, F. A.: Protein handling by the renal tubule. Nephron 13, 22–34 (1974)

Bucher, O., Kaißling, B.: Morphologie des juxtaglomerulären Apparates. Verh. Anat. Ges. *67*, 109–136 (1973)

Bulger, R. E., Nagle, R. B.: Ultrastructure of the interstitium in the rabbit kidney. Am. J. Anat. *136*, 183–204 (1973)

Burg, M. B.: The renal handling of sodium chloride. In: The Kidney. Brenner, B. M., Rector, F. C. (eds.), Vol. I, pp. 272–298. Philadelphia, London, Toronto: Saunders 1976

Burg, M. B., Bourdeau, J. E.: Function of the thick ascending limb of Henle's loop. In: New aspects of renal function (Workshop conferences Hoechst). Vogel, H. G., Ullrich, K. J. (eds.), Vol. VI, pp. 91–102. Amsterdam, Oxford: Excerpta Medica 1978

Burg, M. B., Green, N.: Function of the thick ascending limb of Henle's loop. Am. J. Physiol. *224*, 659–668 (1973)

Burg, M. B., Green, N.: Role of monovalentions in the reabsorption of fluid by isolated perfused proximal renal tubules of the rabbit. Kidney Int. *10*, 221–228 (1976)

Burg, M. B., Orloff, J.: Control of fluid absorption in the renal proximal tubule. J. Clin. Invest. *47*, 2016–2024 (1968)

Burg, M. B., Stoner, L.: Sodium transport in the distal nephron. Fed. Proc. *33*, 31–36 (1974)

Burg, M. B., Grantham, J. J., Abramow, M., Orloff, J.: Preparation and study of fragments of single rabbit nephrons. Am. J. Physiol. *210*, 1293–1298 (1966)

Burg, M. B., Stoner, L. Cardinai, J., Green, N.: Furosemide effect on isolated perfused tubules. Am. J. Physiol. *225*, 119–124 (1973)

Chabardès, D., Imbert, M., Clique, A., Montégut, M., Morel, F.: PTH sensitive adenyl cyclase activity in different segments of the rabbit nephron. Pflügers Arch. *354*, 229–239 (1975 a)

Chabardès, D., Imbert-Teboul, M., Montégut, M., Clique, A., Morel, F.: Catecholamine sensitive adenylate cyclase activity in different segments of the rabbit nephron. Pflügers Arch. *361*, 9–15 (1975 b)

Chabardès, D., Imbert-Teboul, M., Gagnan-Brunette, M., Morel, F.: Different hormonal target sites along the mouse and rabbit nephrons. In: 4th Intern. Sympos. of Biochem. Aspects of kidney function. Schmidt, U., Dubach, U. (eds.). Bern: Huber 1978

Clark, S. L.: Cellular differentiation on the kidneys of newborn mice studied with the electron microscope. J. Biophysic. Biochem. Cytol. *3*, 349–361 (1957)

Costanzo, L., Windhager, E., Taylor, A.: Sodium-calcium interaction in the distal tubule. In: New aspects of renal function. (Workshop conferences Hoechst). Vogel, H. G., Ullrich, K. J. (eds.), Vol. VI, pp. 147–155. Amsterdam, Oxford: Excerpta Medica 1978

Crayen, M., Thoenes, W.: Architektur und cytologische Charakterisierung des distalen Tubulus der Rattenniere. Fortschr. Zool. *23*, 279–288 (1975)

Darnton, S. J.: A possible correlation between ultrastructure and function in the thin descending and ascending limbs of the loop of Henle of rabbit kidney. Z. Zellforsch. *93*, 516–524 (1969)

Deetjen, P., Greger, R., Lang, F.: Renal elimination of endogenous organic adics. In: New aspects of renal function (Workshop conferences Hoechst). Vogel, H. G., Ullrich, K. J. (eds), Vol. VI, pp. 51–53. Amsterdam, Oxford: Excerpta Medica 1978

Di Bona, G.: Neurogenic regulation of renal tubular sodium reabsorption. Am. J. Physiol. *233*, F73–81 (1977)

Dieterich, H. J.: Electron microscopic studies of the innervation of the rat kidney. Z. Anat. Entwickl.-Gesch. *145*, 169–186 (1974)

Dieterich, H. J.: Die Struktur der Blutgefäße in der Rattenniere. Norm. Pathol. Anat. (Stuttgart) *35* (1978)

Dieterich, H. J., Kriz, W.: Interstitium und Lymphgefäße in der Säugerniere. In: Pyelonephritis. Losse, H., Kienitz, M. (eds.), Vol. III, pp. 1–15. Stuttgart: Thieme 1972

Dieterich, H. J., Barrett, J. M., Kriz, W., Bülhoff, J. P.: The ultrastructure of the thin loop limbs of the mouse kidney. Anat. Embryol. *147*, 1–18 (1975)

Diezi, J., Michoud, P., Aceves, J., Giebisch, G.: Micropuncture study of electrolyte transport across papillary collecting duct of the rat. Am. J. Physiol. *224*, 623–634 (1973)

Doležel, S.: The connective tissue skeleton in the mammalian kidney and its innervation. Acta anat. *93*, 194–209 (1975)

Doležel, S, Edvinsson, L., Owman, Ch., Owman, T.: Fluorescene Histochemistry and autoradiography of adrenergic nerves in the renal juxtaglomerular complex of mammals and man, with special regard to the efferent arteriole. Cell. Tissue Res. *169*, 211–220 (1976)

Dousa, T. P., Valtin, H.: Cellular actions of vasopressin in the mammalian kidney. Kidney Int. *10*, 46–63 (1976)

Ericsson, J. L. E., Trump, B. F.: Electron microscopic studies of the epithelium of the proximal tubule of the rat kidney. I. Lab. Invest. *13*, 1427–1456 (1964)

Ericsson, J. L. E., Trump, B. F.: Electron microscopic studies of the epithelium of the proximal tubule of the rat kidney. III. Lab. Invest. *15*, 1610–1533 (1966)

Ericsson, J. L. E., Trump, B. F., Weibel, G.: Electron microscopic studies of the proximal tubule of the rat kidney. II. Lab. Invest. *14*, 1341–1365 (1965)

Ernst, S. A.: Transport ATPase cytochemistry: Ultrastructural localization of potassium-dependent and potassium-independent phosphatase activities in rat kidney cortex. J. Cell Biol. *66*, 586–608 (1975)

Faarup, P.: On the morphology of the juxtaglomerular apparatus. Acta anat. (Basel) *60*, 20–38 (1965)

Faarup, P.: Morphological aspects of the renin-angiotensin system. Copenhagen: Bogtrykkeriet Forum 1971

Fourman, J., Moffat, D. B.: The blood vessels of the kidney. Oxford, Edinburgh: Blackwell Scientific 1971

Frindt, G., Burg, M.: Effect of vasopressin on sodium transport in renal collecting tubules, Kidney Int. *1*, 224–231 (1974)

Frömter, E., Geßner, K.: Free flow potential profile along rat kidney proximal tubule. Pflügers Arch. *351*, 69 83 (1974 a)

Frömter, E., Geßner, K.: Active transport potentials. Membrane diffusion potentials and streaming potentials across rat kidney proximal tubule. Pflügers Arch. *351*, 85–98 (1974b)

Ganote, C. E., Grantham, J. J., Moses, H. L., Burg, M. B., Orloff, J.: Ultrastructural studies of vasopressin effect on isolated perfused renal collecting tubules of the rabbit. J. Cell Biol. *36*, 355–367 (1968)

Giebisch, G.: The distal tubular potassium transport system. New aspects of renal function (Workshop conferences Hoechst). Vogel, H. G., Ullrich, K. J. (eds.), Vol. VI, pp. 136–146. Amsterdam, Oxford: Excerpta Medica 1978

Giebisch, G., Windhager, E. E.: Electrolyte transport across renal tubular membranes. In: Handbook of physiology. Orloff, J., Berliner, R. W., Geiger, S. (eds.), Vol. VIII, pp. 315–376. Washington: Am. Physiol. Soc. 1973

Gorgas, K.: Structure and innervation of the juxtaglomerular apparatus of the rat. Adv. Anat. Embryol. *54*, 5–84 (1978)

Gosling, J. A.: Observations on the distribution of intrarenal nervous tissue. Anat. Rec. *163*, 81–88 (1969)

Gosling, J. A.: Dixon, J. S.: The fine structure of the vasa recta and associated nerves in the rabbit kidney. Anat. Rec. *165*, 503–504 (1969)

Grantham. J. J., Burg, M. B.: Effect of vasopressin and cyclic AMP on permeability of isolated collecting tubules. Am. J. Physiol. *211*, 255–259 (1966)

Grantham, J. J., Irish III, J. M.: Organic acid transport and fluid secretion in the pars recta (PST) of the proximal tubule. In: New aspects of renal function (Workshop conferences Hoechst). Vogel, H.G., Ullrich, K.J. (eds.), Vol. VI, pp. 83-87. Amsterdam, Oxford: Excerpta Medica 1978

Grantham, J. J., Ganote, C. E., Burg, M.B., Orloff, J.: Paths of transtubular water flow in isolated renal collecting tubules. J. Cell Biol. *41*, 562–576 (1969)

Grantham, J. J., Burg, M. B., Orloff, J.: The nature of transtubular Na and K transport in isolated rabbit renal collecting tubules. J. clin. Invest. *49*, 1315–1826 (1970)

Greger, R., Lang, F., Marchand, G., Knox, F. G.: Site of renal phosphate reabsorption. Micropuncture and microinfusion study. Pflügers Arch. *369*, 111–118 (1977)

Griffith, L. D., Bulger, R. E., Trump, B. F.: Structure and staining of mucosubstance on "intercalated cells" from the rat distal convoluted tubule and collecting duct. Anat. Rec. *160*, 643–662 (1968)

Gross, J. B., Imai, M., Kokko, J. P.: A functional comparison of the cortical collecting tubule and the distal convoluted tubule. J. clin. Invest. *55*, 1284–1294 (1975)

Hagège, J., Richet, G.: Dark cells of the distal convoluted tubules and collecting ducts. I. Morphological data. Fortschr. Zool. *23*, 289–298 (1975)

Hagège, J., Gabe, M., Richet, G.: Scanning of the apical pole of distal tubular cells under differing acid-base conditions. Kidney Int. *5*, 137–146 (1974)

Hanker, J. S., Silverman, M. S., Romanovicz, D. K.: Catalase in salivary gland striated and excretory duct cells. II. ∅ Body: an ellipsoidal peroxisomal organelle with crystalloid axial projections. Histochem. J. *9*, 729–744 (1977)

Hays, R. M.: Antidiuretic hormone and water transfer. Kidney Int. *9*, 223–230 (1976)

Helman, S. I., Grantham, J. J., Burg, M. B.: Effect of vasopressin on electrical resistance of renal cortical collecting tubules. Am. J. Physiol. *220*, 1825–1832 (1971)

Hierholzer, K., Wiederholt, M.: Some aspects of distal tubular solute and water transport. Kidney Int. *9*, 198–213 (1976)

Hilger, H. H., Klümper, J. D., Ullrich, K. J.: Wasserrückresorption und Ionentransport durch die Sammelrohrzellen der Säugetierniere. Pflügers Arch. *267*, 218–237 (1958)

Horster, M., Thurau, K.: Micropuncture studies on the filtration rate of single superficial and juxtamedullary glomeruli in the rat kidney. Arch. Ges. Physiol. *301*, 161–181 (1968)

Humbert, F., Pricam, C., Perrelet, A., Orci, L.: Freeze-fracture differences between plasma membranes of descending and ascending branches of the rat Henle's thin loop. Lab. Invest. *33*, 407–411 (1975)

Imai, M.: Function of the thin ascending limb of Henle of rats and hamsters perfused in vitro. Am. J. Physiol. *3*, F 201–F 209 (1977)

Imai, M., Kokko, J. P.: Sodium chloride, urea and water transport in the thin ascending limb of Henle: Generation of osmotic gradients by passive diffusion of solutes. J. clin. Invest. *53*, 393–402 (1974)

Imbert, M., de Rouffignac, C.: Role of sodium and urea in the renal concentrating mechanism in psammomys obesus. Pflügers Arch. *361*, 107–114 (1976)

Imbert, M., Chabardès, D., Montégut, M., Clique, A., Morel, F.: Adenylate cyclase activity along the rabbit nephron as measured in single isolated segments. Pflügers Arch. *354*, 213–228 (1975 a)

Imbert, M., Chabardès, D., Montégut, M., Clique, A., Morel, F.: Vasopressin dependent adenylate cyclase in single segments of rabbit kidney tubule. Pflügers Arch. *357*, 173–186 (1975 b)

Ito, S. pers. comm. (1976)

Iyengar, R., Lepper, K. G., Mailman, D. S.: Involvement of microtubules and microfilaments in the action of vasopressin in canine renal medulla. J. Supramolec. Struct. *5*, 521–530 (1976)

Jacobsen, N. O.: Enzyme histochemical observations on the segmentation of the proximal tubules in the kidney of the female rat. Histochem. *43*, 11–32 (1975)

Jamison, R. L.: Countercurrent systems. In: Kidney and urinary tract physiology. Thurau, K. (ed.), Vol. VI, pp. 199–245. London: Butterworths; Baltimore: University Park Press 1974

Jamison, R. L.: Urinary concentration and dilution. In: The kidney. Brenner, B. M., Rector, F. C. (eds.), Vol. XI, pp. 391–441. Philadelphia, London, Toronto: Saunders 1976

Jamison, R. L., Buerkert, J., Lacy, F. B.: A micropuncture study of collecting tubule function in rats with hereditary diabetes insipidus. J. Clin. Invest. *50*, 2444–2452 (1971)

Jamison, R. L., Lacy, F. B., Pennell, J. P., Sanjana, V. M.: Potassium secretion by the descending limb or pars recta of the juxtamedullary nephron in vivo. Kidney Int. *9*, 323–332 (1976)

Johnston, C. I., Matthews, P. G., Davis, J. M., Morgan, T.: Renin mesaurement in blood collected from the efferent arteriole of the kidney of the rat. Pflügers. Arch. *356*, 277–286 (1975)

Källskog, Ö., Lindbom, L.-O., Ulfendahl, H. R., Wolgast, M.: Hydrostatic pressures within the vascular structures of the rat kidney. Pflügers Arch. *363*, 205–210 (1976)

Kaißling, B., Peter, St., Kriz, W.: The transition of the thick ascending limb of Henle's loop into the distal convoluted tubule in the nephron of the rat kidney. Cell Tissue Res. *182*, 111–118 (1977)

Kaißling, B., Rouffignac, C. de, Barrett, J. M., Kriz, W.: The structural organization of the kidney of the desert rodent psammomys obesus. Anat. Embryol. *148*, 121–143 (1975)

Kawamura, S., Kokko, J.P.: Urea secretion by the straight segment of the proximal tubule. J. Clin. Invest. *58*, 604–612 (1976)

Kazimierczak, J.: Histochemical study of oxidative enzymes in rabbit kidney before and after birth. Acta anat. *55*, 352–369 (1963)

Knepper, M. A., Danielson, R. A., Saidel, C. M., Post, R. S.: Quantitative analysis of renal medullary anatomy in rats and rabbits. Kidney Int. *12*, 313–323 (1977)

116

Knox, F. G.: Regulation of collecting duct sodium reabsorbtion. In: New aspects of renal function (Workshop conferences Hoechst). Vogel, H. G., Ullrich, K. J. (eds.), Vol. VI, pp. 181–183. Amsterdam, Oxford: Excerpta Medica, 1978

Knox, F. G., Haas, J. A., Berndt, T., Marchand, G. R., Youngberg, S. P.: Phosphate transport in superficial and deep nephrons in phosphate-loaded rats. Am. J. Physiol. *233*, F150–F153 (1977)

Kokko, J. P.: Sodium chloride and water transport in the descending limb of Henle. J. Clin. Invest. *49*, 1838–1846 (1970)

Kokko, J. P.: Urea transport in the proximal tubule and the descending limb of Henle. J. Clin. Invest. *51*, 1999–2008 (1972)

Kokko, J. P.: Membrane characteristics governing salt and water transport in the loop of Henle. Fed. Proc. *33*, 25–30 (1974)

Kokko, J. P., Rector, F. C.: Countercurrent multiplication system without active transport in inner medulla. Kidney Int. *2*, 214–223 (1972)

Kokko, J. P., Stokes, J. B., Hanley, M. J., Schwartz, M. J.: Hormonal control of salt transport across the distal nephron. In: New aspects of renal function (Workshop conferences Hoechst). Vogel, H. G., Ullrich, K. J. (eds.), Vol. VI, pp. 129–135. Amsterdam, Oxford: Excerpta Medica 1978

Kriz, W.: Der architektonische und funktionelle Aufbau der Rattenniere. Zellforsch, Z. *82*, 495–535 (1967)

Kriz, W.: Organization of structures within the renal medulla. Proceedings of an international colloquy held at Sarasota, (Florida) 1968. In: Urea and the Kidney. Schmidt-Nielsen, B., Kerr, D. W. S. (eds.), pp. 342–357. Amsterdam: Excerpta Medica 1970

Kriz, W., Dieterich, H. J.: Das Lymphgefäßsystem bei einigen Säugetieren. Licht- und elektronenmikroskopische Untersuchungen. Z. Anat. Entwickl.-Gesch. *131*, 111–147 (1970)

Kriz, W., Koepsell, H.: The structural organization of the mouse kidney. Z. Anat. Entwickl.-Gesch. *144*, 137–163 (1974)

Kriz, W., Schnermann, J., Koepsell, H.: The position of short and long loops of Henle in the rat kidney. Z. Anat. Entwickl.-Gesch. *138*, 301–319 (1972 a)

Kriz, W., Schnerman, J., Dieterich, H. J.: Differences in the morphology of descending limbs of short and long loops of Henle in the rat kidney. In: Recent advances in renal physiology. Wirz, H., Spinelli, F. (eds.), pp. 140–144. Basel: Karger 1972 b

Kriz, W., Barrett, J. M., Peter, S.: The renal vasculature: Anatomical-functional aspects. In: Kidney and urinary Tract Physiology II. Thurau, K. (ed.), Vol. 11, pp. 1–21. Baltimore, London, Tokyo: Univ. Park Press 1976

Kriz, W., Kaißling, B., Pszolla, M.: Morphological characterization of the cells in Henle's loop and the distal tubule. In: New aspects of renal function (Workshop conferences Hoechst). Vogel, H. G., Ullrich, K. J. (eds.), Vol. VI, pp. 67–78. Amsterdam, Oxford: Excerpta Medica 1978

Kyte, J.: Immunoferritin determination of the distribution of $(Na^+ + K^+)$ ATPase over the plasma membranes of renal convoluted tubules. J. Cell Biol. *68*, 287–303 (1976)

Laurence, R., Marsh, D. J.: Effect of diuretic states on hamster collecting duct electrical potential differences. Am. J. Physiol. *220*, 1610–1616 (1971)

Latta, H.: Ultrastructure of the glomerulus and juxtaglomerular apparatus. Handbook of Physiology. Orloff, J., Berliner, R. W., Geiger, S. (eds.), Vol. VIII, pp. 1–29. Washington: Am. Physiol. Soc. 1973

Lee, S. H.: The possible role of the vesicles in renal ammonia excretion. Cell. Biol. *45*, 644–649 (1970)

Lönnerholm, G.: Histochemical demonstration of carbonic anhydrase activity in the rat kidney. Acta Physiol. Scand. *81*, 433–439 (1971)

Lönnerholm, G.: Histochemical demonstration of carbonic anhydrase activity in the human kidney. Acta Physiol. Scand. *88*, 455–468 (1973)

Mc Donald, K. M., Miller, P. D., Anderson, R. J., Berl, T., Schrier, R. W.: Hormonal control of renal water excretion. Kidney Int. *10*, 38–45 (1976)

Maunsbach, A. B.: Observations on the segmentation of the proximal tubule in the rat kidney. J. Ultrastruct. Res. *16*, 239–258 (1966)

Maunsbach, A. B.: Ultrastructure of the proximal tubule. In: Handbook of Physiology. Orloff, J., Berliner, R. W., Geiger, S. (eds.), Vol. *VIII*, pp. 31–79. Washington: Am. Physiol. Soc. 1973

Möllendorff, W. v.: Der Exkretionsapparat. In: Handbuch der mikroskopischen Anatomie. Möllendorff, W. v. (ed.), Vol. *VII/1*, pp. 1–327. Berlin: Springer 1930

Morel. F., Chabardes, D., Imbert, M.: Functional segmentation of the rabbit distal tubule by microdetermination of hormone-dependent adenylate cyclase activity. Kidney Int. *9*, 264–277 (1976)

Müller, J., Barajas, L.: Electron microscopic and histochemical evidence for a tubular innervation in the renal cortex of the monkey. J. Ultrastruct. Res. *41*, 533–549 (1972)

Myers, C. E., Bulger, R. E., Tisher, C. C., Trump, B. F.: Human renal ultrastructure. IV. Collecting duct of healthy individuals. Lab. Invest. *15*, 1921–1950 (1966)

Norvell, J. E.: A histochemical study of the adrenergic and cholinergic innervation of the mammalian kidney. Anat. Rec. *163*, 236 (1969)

Novikoff, A. B., Novikoff, P. M.: Microperoxisomes. J.Histochem.Cytochem. *21*, 963–966 (1973)

Novikoff, A. B., Novikoff, P. M., Davis, C., Quintana, N.: Studies on microperoxisomes. II. A cytochemical method for light and electron microscopy. J. Histochem. Cytochem. *20*, 1006–1023 (1972)

O'Morchoe, C. C. C., O'Morchoe, P. J., Helmes, M. J., Jarosz, H. M.: Flow and composition of renal hilar lymph during volume expansion and saline diuresis. Lymphology *11*, 27–31 (1978)

Oliver, J.: Nephrons and kidneys. New York, Evanston, London: Harper & Row, Hoeber Medical Division 1968

Osathanodh, V., Potter, E. L.: Development of human kidney as shown by microdissection. III. Formation and interrelationship of collecting tubules and nephrons. Arch. Pathol. Lab. Med. *76*, 290–302 (1963)

Orstavik, T. B., Nustad, K., Brandtzaeg, P., Pierce, J. V.: Cellular origin of urinary kallikreins. J. Histochem. Cytochem. *24*, 1037–1039 (1976)

Osvaldo-Decima, L.: Ultrastructure of the lower nephron. Handbook of Physiology. Orloff, L., Berliner, R. W., Geiger, S. (eds.), Vol, *VIII*, pp. 81–102. Washington: Am. Physiol. Soc. 1973

Peter, K.: Untersuchungen über Bau und Entwicklung der Niere. Jena: Gustav Fischer, Vol. *1*: 1909; Vol. *2*: 1927

Pfaller, W., Rittinger, M.: Quantitative Morphologie der Niere. Mikroskopie *33*, 74–79 (1977)

Pfeiffer, E. W.: Comparative anatomical observations of the mammalian renal pelvis and medulla. J. Anat. (Lond.) *102*, 321–331 (1968)

Pinter, G. G.: Pers. comm (1977)

Potter, E. L.: Normal and abnormal development of the kidney. Chicago: Year Book Medical Publishers 1972

Poujeol, P., Chabardès, D., Roinel, N., Rouffignac, C. de: Influence of extracellular fluid volume expansion on magnesium, calcium and phosphate handling along the rat nephron. Pflügers Arch. *365*, 203–211 (1976)

Poujeol, P., Corman, B., Touvay, C., Rouffignac, C. de: Phosphate reabsorption in rat nephron terminal segments. Intrarenal heterogenity and strain differences. Pflügers. Arch. *371*, 39–44 (1977)

Pricam, C., Humbert, F., Perrelet, A., Orci, L.: A freeze-etch study of the tight junctions of the rat kidney tubules. Lab. Invest. *30*, 286–291 (1974)

Rhodin, J.: Anatomy of kidney tubules. Int. Rev. Cytol. *VII*, 485–534 (1958)

Richardson, K. C., Jarret, L., Finke, E. H.: Embedding in epoxy resins for ultrathin sectioning in electromicroscopy. Stain Technol. *35*, 313–323 (1960)

Richet, G., Hagège, J.: Dark cells of the distal convoluted tubules and collecting ducts. II. Physiological significance. Fortschr. Zool. *23*, 299–306 (1975)

Richet, G., Hagège, J., Gabe, M.: Corrélations entre les transferts de bicarbonate et la morphologie du segment terminal du néphron chez le rat. Nephron *7*, 413–429 (1970)

Robertson, G. L., Shelton, R. L., Athar, S.: The osmoregulation of vasopressin. Kidney Int. *10*, 25–37 (1976)

Rocha, A. S., Kokko, J. H.: Sodium chloride and water transport in the medullary thick ascending limb of Henle. Evidence for active chloride transport. J. clin. Invest. *52*, 612–623 (1973)

Rocha, A. S., Kokko, J. P.: Permeability of medullary nephron segments to urea and water: Effect of vasopressin. Kidney Int. *6*, 379–387 (1974)

Rocha, A. S., Magaldi, J. B., Kokko, J. P.: Calcium and phosphate transport in isolated segments of rabbit Henle's loop. J. Clin. Invest. *59*, 975–983 (1977)

Roch-Ramel, F., Filloux, B., Guignard, J. P., Peters, C.: Fate of urea in Henle's loops of the rabbit and the rat. In: New aspects of renal function (Workshop conferences Hoechst). Vogel, H. G., Ullrich, K. J. (eds.), Vol. *VI*, pp. 118–121. Amsterdam, Oxford: Excerpta Medica 1978

Roesinger, B., Schiller, A., Taugner, R.: A freeze-fracture study of tight junctions in the pars convoluta and pars recta of the renal proximal tubule. Cell Tissue Res. *186*, 121–133 (1978)

Rollhäuser, H.: Untersuchungen über den örtlichen und zeitlichen Ablauf der Phenolrot-Ausscheidung in den Tubuli der unbeeinflußten Rattenniere. Zellforsch, Z. *51*, 348–355 (1960)

Rollhäuser, H., Kriz, W. Heinke, W.: Das Gefäßsystem der Rattenniere. Zellforsch, Z. *64*, 381–403 (1964)

Rosen, S.: Localization of carbonic anhydrase activity in the vertebrate nephron, Histochem. J. *4*, 35–48 (1972)

Rouffignac, C. de, Deiss, S., Bonvalet, J. P.: Détermination du taux individuel de filtration glomérulaire des néphrons accessibles et inaccessibles a la microponction. Pflügers Arch. *315*, 273–290 (1970)

Rouffignac, C. de, Morel, F., Moss, N., Roinel, N.: Micropuncture study of water and electrolyte movement along the loop of Henle in Psammomys with special reference to magnesium, calcium and phosphorus. Pflügers Arch. *344*, 309–326 (1973)

Rytand, D. A.: The number and size of mammalian glomeruli as related to kidney and to body weight with method for the enumeration and measurement. Am. J. Anat. *62*, 507–520 (1938)

Sakaguchi, H., Suzuki, Y.: Fine structure of renal tubule cells. Keio J. Med. *7*, 17–26 (1958)

Schafer, J. A., Andreoli, T. E.: The effect of antidiuretic hormone on solute flows in mammalian collecting tubules. J. Clin. Invest. *51*, 1279–1286 (1972)

Schafer, J. A., Patlak, C. S., Andreoli, T. E.: Fluid absorption and active and passive ion flows in the rabbit superficial pars recta. Am. J. Physiol. *233*, F154–F167 (1977)

Schiebler, T. H.: Morphologie der Nieren und ihrer Ableitungswege. In: Handbuch der Zoologie. Helmcke, J. G., Lengerhen, H. v., Stark, D. (eds.) Vol. *VIII*, pp. 1–84. Berlin: De Gruyter 1959

Schiller, A., Taugner, R., Roesinger, B.: Vergleichende Morphologie der zonulae occludentes am Nierentubulus. Verh. Anat. Ges. *72*, 229–234 (1978)

Schmidt, U., Dubach, U. C.: Activity of (Na^+ K^+)-stimulated adenosintriphosphatase in the rat nephron. Pflügers Arch. *306*, 219–226 (1969)

Schmidt, U., Dubach, U. C.: Quantitative Histochemie am Nephron. Progr. Histochem. Cytochem. *2*, 185–298 (1971)

Schmidt, U., Schmid, J., Schmid, H., Dubach, U. C.: Sodium- and potassium-activated ATPase. A possible target of aldosterone. J. Clin. Invest. *55*, 655–660 (1975)

Schmidt-Nielsen, B.: Excretion in mammals: role of the renal pelvis in the modification of the urinary concentration and composition. Fed. Proc. *36*, 2493–2503 (1977)

Schnermann, J.: The role of the juxtaglomerular apparatus in single nephron function. In: New aspects of renal function (Workshop conferences Hoechst). Vogel, H. G., Ullrich, K. J. (eds)., Vol. *VI*, pp. 189–197. Amsterdam, Oxford: Excerpta Medica 1978

Schnermann, J., Ploth, D. W., Hermle, M.: Activation of tubulo-glomerular feedback by chloride transport. Pflügers Arch. *362*, 229–240 (1976)

Schønheyder, H. C., Maunsbach, A. B.: Ultrastructure of a specialized neck region in the rabbit nephron. Kidney Int. *7*, 145–153 (1975)

Schwartz, M. M., Venkatachalam, M. A.: Structural differences in thin limbs of Henle: Physiological implications. Kidney Int. *6*, 103–208 (1974)

Sheehan, H. L., Davis, J. C.: Anatomy of the pelvis in the rabbit kidney. J. Anat. *93*, 499–502 (1959)

Smith, H. W.: The kidney structure and function in health and disease. New York: Oxford University Press 1951

Sonnenberg, H.: Medullary collecting-duct function in adiuretic and in salt – or water – diuretic rats. Am. J. Physiol. *226*, 501–506 (1976)

Sperber, J.: Studies on the mammalian kidney. Zool. Bidrag (Uppsala) *22*, 249–431 (1944)

Stein, J. H., Osgood, R. W., Kunau, R. T.: Direct measurement of papillary collecting duct sodium transport in the rat. J. Clin. Invest. *58*, 767–773 (1976)

Stephenson, J. L.: Concentrating engines and the kidney. I. Central core model of the renal medulla. Biophys. J. *13*, 512–445 (1973a)

Stephenson, J. L.: Concentrating engines and the kidney. II. Multisolute central core system. Biophys. J. *13*, 546–567 (1973b)

Stephenson, J. L.: Concentration of urine in a central core model of the renal counterflow system. Kidney Int. *2*, 85–94 (1972)

Stephenson, J. L., Tewarson, R. P., Mejia, R.: Quantitative analysis of mass and energy balance in non-ideal models of the renal counterflow system. Proc. Natl. Acad. Sci. USA *71*, 1618–1622 (1974)

Swann, H. G., Norman, R. J.: The periarterial spaces of the kidney. Texas Rep. Biol. Med. *28*, 318–335 (1970)

Thoenes, W.: Die Mikromorphologie des Nephrons in ihrer Beziehung zur Funktion. I. Funktionseinheit: Glomerulum – proximales und distales Konvolut. Klin. Wochenschr. *39*, 504–518 (1961a)

Thoenes, W.: Die Mikromorphologie des Nephrons in ihrer Beziehung zur Funktion. II. Funktionseinheit: Henlesche Schleife – Sammelrohr. Klin. Wochenschr. *39*, 827–839 (1961b)

Thoenes, W., Langer, K. H.: Relationship between cell structures of renal tubules and transport mechanisms. In: Renal transport and diuretics. Thurau, K., Gahrmärker, H. (eds.), pp. 37–64. Berlin, Heidelberg, New York: Springer 1969

Tisher, C. C.: Anatomy of the kidney. In: The kidney. Brenner, B. M., Rector, F. C. (eds.), Vol. I, pp. 3–64. Philadelphia, London, Toronto: Saunders 1976

Tisher, C. C., Yarger, W. E.: Lanthanum permeability of the tight junctions (zonula occludens) in the renal tubule of the rat. Kidney Int. *3*, 238–250 (1973)

Tisher, C. C., Yarger, W. E.: Lanthanum permeability of tight junctions along the collecting duct of the rat. Kidney Int. *7*, 35–44 (1975)

Tisher, C. C., Bulger, R. E., Trump, B. F.: Human renal ultrastructure. I. Proximal tubule of healthy individuals. Lab. Invest. *15*, 1357–1394 (1966)

Tisher, C. C., Bulger, R. E., Trump, B. F.: Human renal ultrastructure. III. The distal tubule in healthy individuals. Lab. Invest. *18*, 655–668 (1968)

Tisher, C. C., Bulger, R. E., Valtin, H.: Morphology of renal medulla in water diuresis and vasopressin-induced antidiuresis. Am. J. Physiol. *220*, 87–94 (1971)

Toback, F. G., Ordonez, N. G., Bortz, S. L., Spargo, B. H.: Zonal changes in renal structure and phospholipid metabolism in potassium-deficient rats. Lab. Invest. *34*, 115–124 (1976)

Trueta, J., Bareley, A. E. Daniel, P. M., Franklin, K. J., Prichard, M. M. L.: Studies of the renal circulation. Vol. *VI*, pp. 39–90. Springfield (Ill.): Thomas 1947

Tune, B., Burg, M.: Glucose transport by proximal tubules. Am. J. Physiol. *221*, 580–585 (1971)

Tune, B., Burg, M., Patlak, C.: Characteristics of p-aminohippurate transport in proximal renal tubules. Am. J. Physiol. *217*, 1057–1063 (1969)

Uhlich, E., Baldamus, C. A., Ullrich, K. J.: Einfluss von Aldosteron auf den Natriumtransport in den Sammelrohren der Säugetierniere. Arch. Ges. Physiol. *308*, 111–126 (1969)

Ullrich, K. J., Rumrich, G., Klöss, S.: Phosphate transport in the proximal convolution of the rat kidney. I. Tubular heterogeneity, effect of parathyroid hormone in acute and chronic parathyroidectomized animals and effect of phosphate diet. Pflügers Arch. *372*, 269–274 (1977)

Valtin, H.: Structural and functional heterogeneity of mammalian nephrons. Am. J. Physiol. *233*, F491–F501 (1977)

Vogel, G., Gärtner, K., Ulbrich, M.: The flow rate and macromolecule content of hilar lymph from the rabbit's kidney under conditions of renal venous pressure elevation and restriction of renal function. Studies on the origin of renal lymph. Lymphology 7, 136–143 (1974)

Welling, L. W., Welling, D. J.: Surface areas of brush border and lateral cell walls in the rabbit proximal nephron. Kidney Int. *8*, 343–348 (1975)

Welling, L. W., Welling, D. J.: Shape of epithelial cells and intercellular channels in the rabbit proximal nephron. Kidney Int. *9*, 385–394 (1976)

Woodhall, P. B., Tisher, C. C.: Response of the distal tubule and cortical collecting duct to vasopressin in the rat. J. Clin. Invest. *52*, 3095–3108 (1973)

Wright, F. S.: Sites and mechanisms of potassium transport along the renal tubule. Kidney Int. *11*, 415–432 (1977)

Wright, F. S., Giebisch, G.: Glomerular filtration in single nephrons. Kidney Int. *1*, 201–209 (1972)

Zimmermann, K. W.: Zur Morphologie der Epithelzellen der Säugetierniere. Arch. Mikroskop. Anat. *78*, 199–231 (1911)

Zimmermann, K. W.: Über den Bau des Glomerulus der Säugetierniere. Z. Mikrosk. Anat. Forsch. *32*, 176–278 (1933)

Subject Index

Other Reviews of Interest in this Series

Springer-Verlag Berlin Heidelberg New York

Reviews and critical articles covering the entire field of normal anatomy (cytology, histology, cyto- and histo-chemistry, electron microscopy, macroscopy, experimental morphology and embryology and comparative anatomy) are published in Advances in Anatomy, Embryology and Cell Biology. Papers dealing with anthropology and clinical morphology that aim to encourage co-operation between anatomy and related disciplines will also be accepted. Papers are normally commissioned. Original papers and communications may be submitted and will be considered for publication provided they meet the requirements of a review article and thus fit into the scope of "Advances". English language is preferred, but in exceptional cases French or German papers will be accepted.

It is a fundamental condition that submitted manuscripts have not been and will not simultaneously be submitted or published elsewhere. With the acceptance of a manuscript for publication, the publisher acquires full and exclusive copyright for all languages and countries.

Twenty-five copies of each paper are supplied free of charge.

Manuscripts should be addressed to

Prof. Dr. A. **BRODAL,** Universitetet i Oslo, Anatomisk Institutt, Karl Johans Gate 47 (Domus Media), Oslo 1/Norway

Prof. W. **HILD,** Department of Anatomy, Medical Branch, The University of Texas, Galveston, Texas 77550/USA

Prof. Dr. J. van **LIMBORGH,** Universiteit van Amsterdam, Anatomisch-Embryologisch Laboratorium, Mauritskade 61, Amsterdam-O/Holland

Prof. Dr. R. **ORTMANN,** Anatomisches Institut der Universität, Lindenburg, D-5000 Köln-Lindenthal

Prof. Dr. T. H. **SCHIEBLER,** Anatomisches Institut der Universität, Koellikerstraße 6, D-8700 Würzburg

Prof. Dr. G. **TÖNDURY,** Direktion der Anatomie, Gloriastraße 19, CH-8006 Zürich/Schweiz

Prof. Dr. E. **WOLFF,** Collège de France, Laboratoire d'Embryologie Expérimentale, 11, Place Marcelin Berthelot, F-75005 Paris/France